Evolution and the Naked Truth

A Darwinian approach to philosophy

GONZALO MUNÉVAR

Ashgate

Aldershot • Brookfield USA • Singapore • Sydney

© Gonzalo Munévar 1998

All rights reserved. No part of this publication may be reproduced, stored in a retrieval system, or transmitted in any form or by any means, electronic, mechanical, photocopying, recording or otherwise without the prior permission of the publisher.

Published by
Ashgate Publishing Ltd
Gower House
Croft Road
Aldershot
Hants GU11 3HR
England

Ashgate Publishing Company
Old Post Road
Brookfield
Vermont 05036
USA

British Library Cataloguing in Publication Data
Munévar, Gonzalo
  Evolution and the naked truth: a Darwinian approach to
  philosophy. - (Avebury series in philosophy)
  1. Darwin, Charles, 1809 - 1882 2. Philosophy 3. Evolution
  I. Title
  576.8'01

Library of Congress Catalog Card Number: 98-71956

ISBN 1 84014 344 4

Printed and bound by Athenaeum Press, Ltd.,
Gateshead, Tyne & Wear.

Contents

Acknowledgements vii
Introduction ix

PART I: EVOLUTIONARY RELATIVISM

1 Evolution and the Naked Truth 3

2 Extraterrestrial and Human Science 23

3 Bohr and Evolutionary Relativism 33

4 Cultural Relativism and Universalism 45

5 A Note on Margolis 57

PART II: EVOLUTION AND RATIONALITY

6. The Connection Between Evolution and the Nature of Scientific Knowledge 65

7 Towards a Future Epistemology of Science 75

8 Hull, Biology, and Epistemology 91

9 Science as Part of Nature 99

PART III: FROM EPISTEMOLOGY TO ETHICS

10 Evolution and Justification 113

| 11 | The Morality of Rational Ants | 131 |
| 12 | A Naturalistic Account of Free Will | 149 |

PART IV: APPLICATION TO SPACE SCIENCE

13	A Philosopher Looks at Space Exploration	169
14	Pecking Orders and Rhetoric in Science	181
15	SETI, Self-Reproducing Machines, and Impossibility Proofs	189

APPENDICES ON FEYERABEND

| A | Science in Feyerabend's Free Society | 199 |
| B | A *Rehabilitation* of Paul Feyerabend | 219 |

Index   245

Acknowledgements

I owe a debt of gratitude to the many people who commented on the early drafts of the papers that have now become chapters of this book. I am particularly grateful to Bill Blizek, Jerry Cederblom, Arthur Diamond, Steve Fuller, Michael Katz, Jesús Mosterín, Andrés Rivadulla, Peter Singer, and E.O. Wilson. Throughout my career I have also been blessed not only with the criticism but the encouragement of Paul Churchland, Clifford Hooker, David Paulsen, Javier Ordoñez, and Sheldon Reaven. Many a conversation with them has ended up in these pages, although I do not wish to make them in any way responsible for my peculiar views. Three people have had a very direct influence on the contents of this book. One is the late Paul Feyerabend, whose work continues to challenge and inspire me, and whose friendship will always be a most cherished memory. Another is Wallace Matson, to whose prompting I owe the writing of "Evolution and Justification" and "The Morality of Rational Ants," and whose kindness and advice have been a comfort and a guide over the years. And I wish to make a special mention of David Lamb, who as a friend and editor has been unfailing in his support, who has motivated me with his understanding of my work, and who has added entertainment to enlightenment with his outrageously funny stories.

I am grateful to the Philosophy Department of the University of California, Irvine, and to Gerasimos Santas, the Chair, for their hospitality during the months I have spent finishing this book. I also thank Joanna Gislason and Martha Sanchez for typing and readying the manuscript, and Pauline Beavers, of Ashgate, for her patient advice.

I wish to thank the following editors and publishers for permission to reprint the papers that form part of this book: E.J. Brill, for "Evolution and the Naked Truth," from M. Dascal's *Cultural Relativism and Philosophy*,1991; *Explorations in Knowledge*, for "Towards a Future Epistemology of Science," Vols. I, II, 1984, "Pecking Orders and the Rhetoric of Science," Vol. III, No. 2, 1986, "Human and Extraterrestrial Science," Vol. VI, No. 2, 1989, "Bohr and Evolutionary Relativism," Vol. XII, No. 2, 1995, and significant portions of "The Morality of Rational Ants," whose first version appeared as a review of Peter Singer's *The Expanding Circle*, in 1987; *Biology and Philosophy*, for "Hull, Biology and Epistemology," Vol. 3, No. 2, 1988; SUNY Press, for "Science as Part of Nature," from K. Hahlweg and C.A. Hooker, eds., *Issues in Evolutionary Epistemology*, 1989; Kluwer Academic Publishers, for "Science in

Feyerabend's Free Society," from G. Munévar, ed., *Beyond Reason: Essays on the Philosophy of Paul K. Feyerabend*, 1991; Iowa State University Press, for "The Connection between Evolution and the Nature of Scientific Knowledge," from G.P. Scott and H.W. Wahner, eds., *Radiation and Cellular Response*, 1983; *Social Epistemology*, for "Comment on Margolis' Pragmatic Realism," Vol. 2, No. 3, 1988; *The Monist*, for "Evolution and Justification," Vol. 71, No. 3, 1988. Four papers are scheduled to appear shortly after this writing, January 1998. I thank the National Space Society, for "A Philosopher Looks at Space Exploration," from *Proceedings of the International Conference on Space Development*; *The Philosophic Exchange*, for SETI, "Self-Reproducing Machines, and Impossibility Proofs"; *Enciclopedia Iberoamericana de Filosofía*, for "Cultural Relativism and Universalism" (translated from the Spanish "Relativismo y Universalismo Culturales"); and *Dialogos*, for "A Naturalistic Account of Free Will."

Introduction

By the time I became an upper-division student I already thought there was something wrongheaded about mainstream philosophy. The emphasis on linguistic or logicist taxonomy, so far divorced from the plain fact (to me) that human beings were part of nature, did not seem all that promising. At times the school philosophizing resembled an attempt to determine the nature of radios strictly from the music that comes out of them. Any questions in the direction suggested by my intuitions were met with the reply that they were empirical questions, and thus presumably not philosophical. Nevertheless, I was generally a good boy, minded my p's and q's, and grew up to be a graduate student. At Berkeley I was fortunate to fall under the tutelage of thinkers who not only encouraged me but even found some of my ideas valuable. Michael Scriven, Hans Sluga, and Wallace Matson come fondly to mind. Thanks to them, and to the relentless criticism from Paul Feyerabend and much wise guidance from Gunther Stent I made it through my Ph.D. and beyond that to my first book, *Radical Knowledge*, the first thorough development of my biological epistemology. Since then I have refined my ideas and expanded them into other areas of philosophy. This book is a selection of my published efforts to produce a Darwinian approach to philosophy.

I have divided the book into four parts and two appendices. Part I deals with two main theses:

> (1) That if we seriously apply evolutionary theory to understand scientific knowledge, relativism is unavoidable. This result is exactly the opposite of the support for realism that Popper and others had hoped for. In arriving at this significant conclusion, it will become evident that evolutionary relativism can easily withstand all the favorite objections against relativism. Worse still, it will be seen that some of the most important traditional objections rest on fallacious reasoning.

> (2) That relativism need not be restricted to a skeptical role. Evolutionary relativism can serve as a positive philosophy thanks to the development of a theory of relative truth. This theory is first sketched in Chapter 1, considered anew in the context of Bohr's interpretation of quantum physics (Chapter 3), and then applied to culture in general, and art in particular, in Chapter 4.

Two short essays complete Part I. Chapter 2 explores the conditions necessary for extraterrestrial science to exist. This exploration is used as a device for understanding the development of our own science. Chapter 5 is a comment on Margolis' internal relativism that allows me to discuss his favoring of Quine's view over those of Kuhn and Feyerabend.

Part II deals with the issue of the rationality of science. My main thesis is that the application of evolutionary theory can help us show that science is a rational activity. My approach is rather literal, more in the tradition of Konrad Lorenz's epistemology than in that of standard evolutionary epistemology. I wish to show that science is part of nature, not that it is like nature. I am interested in determining the biological conditions that make science possible rather than the rationality of the *history* of science (which I think cannot be shown by the application of Darwinian models). To achieve my goal I also make use of a social conception of scientific rationality.

Chapter 6 serves as brief introduction to evolutionary epistemology and as transition from my *Radical Knowledge* to my later work. In Chapter 7 I argue for a prescriptive naturalism. I point out there how the problem of the rationality of science should be framed anew in light of the revolution in the philosophy of science, and then I suggest how this problem can be solved with my biological approach. This theme is developed further in a short note on the work of David Hull (Chapter 8) and in the closing chapter of the section, appropriately titled, I believe, "Science as Part of Nature."

The first chapter of Part III ("Evolution and Justification") serves as the transition from epistemology to ethics, a transition made necessary by my argument, in Chapter 9, to the effect that the prescriptive task of epistemology can be finished only by crossing over into ethics. Chapter 10 ("The Morality of Rational Ants") is entirely devoted to the relevance of biology to ethics. This chapter is an extensively revised version of a review of Peter Singer's *The Expanding Circle*, in which Singer explains and criticizes the attempt by E.O. Wilson to have sociobiology take over ethics. I praise Singer for his explanation but disagree with this criticism. I support Wilson's general approach and argue that the much-revered naturalistic fallacy is a red herring (my rule of thumb is that fallacies with Latin names indeed point to mistakes in reasoning, while fallacies with English names point to mistakes by those who "discovered" them). One of the results from this chapter is the realization that evolutionary relativism is also present in ethics.

I close Part III with "A Naturalistic Account of Free Will," in which I call attention to some serious flaws in Hume's solution of the problem, and in which I argue that a naturalistic account based on the work of Francis Crick, Paul Churchland, and others is compatible with free will.

Part IV is an application of my general philosophy to space science and exploration. For years I have used this field as a testing ground for my view concerning the ultimate practicality of science. Chapter 13 gives an introduction to the philosophy of space exploration. Chapter 14 considers the question of the relationship between pure and applied science in the context of space physics and astronomy. Even though string theory and other views popular around the time when the article was written have been abandoned, the points made in it still illustrate how space physics and astronomy are not exceptions to that ultimate practicality (via their connection with "fundamental" science). I close this part with a critique of John von Neumann's view of biological beings as machines by discussing NASA's use of his work on self-reproducing automata.

The two appendices on Feyerabend are intended to support my views on the rationality of science, but they are also an opportunity to explain his philosophy. Appendix B, in particular, will show how his epistemology blazed new trails for the philosophy of science and why those in the field should continue to pay attention to his writings. Appendix A criticizes his views on the relationship between science and a free society (views which, incidentally, he changed towards the end of his life). It serves to set proper boundaries for some of my remarks in Parts II and III.

I must include two apologies to the reader. The first is that my approach is Darwinian in that I apply evolutionary theory to a variety of philosophical problems. But it is not an examination of Darwin's own views on these problems. I hope I have not mislead anyone. The second is that there is a considerable amount of overlap in some of the chapters. Several attempts to correct this flaw resulted in mangled essays and finally convinced me that the cure was worse than the disease. Each essay was initially written to stand on its own, and I eventually decided to let the reader value each essay on its own merits, with a minimum of changes.

I will end this introduction by expressing what I hope to obtain from the publication of this book. The views I advance in these essays represent efforts to achieve a comprehensive account of humans as part of nature. I would like to think that I have made a good start in that direction by clearing from my path many philosophical obstacles. But the road still ahead of me stretches far into the distance, and thus I hope to receive the kind of critical commentary that will help me recognize the pitfalls that surely await me.

PART I
EVOLUTIONARY RELATIVISM

1 Evolution and the Naked Truth*

In this paper I wish to defend an evolutionary form of epistemological relativism. I will first explain why evolutionary considerations lead to a complete relativism (perceptual, intellectual, scientific). I will then defend this evolutionary relativism from standard and new objections advanced by realists. And I finally will sketch an evolutionary theory of relative truth. As I will indicate, evolutionary relativism will support at least some forms of social relativism.

I. Social Relativism and the Naked Truth

There may be several types of suspicions cast upon social relativism, depending on the realist theses we favor. We may, for example, hold that social relativism, as all forms of relativism, must be ultimately incoherent. This suspicion may remain, even in the presence of strong reasons in favor of social relativism, because an entrenched realism would lead us to assume that either the truth of the matter has not yet been discovered in those areas in question, or else that there is no truth of the matter. Of course, a relativist may also hold that there is no truth of the matter in those areas. The difference between him and the realist is that the realist would think that those areas where he makes relativist concessions are somehow not as worthy, epistemologically speaking, as those areas where knowledge is really possible. In this regard, the tradition used to contrast natural science with soft endeavors such as ethics, art, or even the social sciences. After the work of Kuhn (1970), Feyerabend (1975), and others, the matter has become far more complicated, for it appears that science itself is subject to the vagaries of social relativism. A despairing realist may conclude that there is no truth of the matter in any empirical investigation, but I suspect that most realists who can recognize the social relativization of natural science would simply fall back on the position that the truth of the matter has not been found yet. The realist need not expect that the truth of the matter will ever be found, although he likes to say that science gets us closer and closer to it. All the realist needs is the belief that the truth of the matter indeed exists. The truth may be perceived disguised by social clothing, but underneath that clothing it stands naked in all its glory. The realist, like adolescents of all ages, prizes above all, the uncovering of nakedness.

This realist belief and the consequent attitude, often referred to jointly as "metaphysical realism," raise a choir of pious lamentations from many quarters. For some, it makes no sense to speak of a truth to which we cannot aspire. What bothers these philosophical puritans is the notion of a truth which is in principle impossible. But it seems to me that the metaphysical realist may hold that the naked truth is only practically, not in principle, unattainable. Or, when confronted with arguments to the effect that the truth must always appear clothed to us, he may try some of the sophisticated answers that we will examine later in this paper. In any event, the underlying reason for disowning metaphysical realism is that since we may never know the naked truth we cannot really decide whether there is such a thing. A respectable philosopher does not want to be caught holding positions based on undecidable assumptions. I suspect that this sense of philosophical propriety leads to some very confused forms of realism. Be that as it may, however, its starting point is wrong: the question of the naked truth is decidable. Indeed, metaphysical realism is false! Truth, like beauty, is in the eye of the beholder.

II. Evolutionary Relativism

Once we come to understand the philosophical strength of epistemological relativism, we will be able to shed some light on some murky issues in the background of social relativism. I intend to argue, then, against the epistemological ideal offered by realism: to come to know the naked truth, or rather, to come to know the way things really are (in epistemology: to come to know *the* structure of the world). It is supposed that humans will fall short because their senses are prone to distortion and their intellects to prejudice. I will argue that there is no such thing as knowing the ways things really are. Absolute knowledge is a mistake even as an ideal. The relativism I have in mind will grow out of evolutionary considerations on the nature of knowledge. Popper and others have thought that evolutionary theory would in some way provide a warrant for realism (an evolutionary variation of "science is successful because it approximates the truth"). But as we will see, careful attention to the implications of evolution will turn the realist dictum on its head.

1. Natural History and Knowledge

Let me begin very much in the spirit of a committed realist. Science is produced by human beings; by beings, that is, that use their brains in social cooperation. Those brains are the result of a long evolutionary history, as is the inclination to form groups in order to solve a variety of problems. It is plausible to suppose, then, that the capacity to know, and to organize socially in order to know, may have some biological basis. And therefore, it would seem that biology could offer some guidance in understanding the nature of empirical knowledge. We notice, for example, that the sort of empirical knowledge possessed by organisms, at least at an elementary level, is largely the result of the interactions between those organisms and their world; or to be more specific, a result of the interaction between the biology of the organisms and their environment. A bird, say, perceives as it does because natural selection has produced its kind of perceptual apparatus, and because that perceptual apparatus has been fine-tuned to a range of environments through the bird's individual development. Its perception is thus the result of two histories: (1) a long history of interactions between its ancestors and a sequence of environments, which have produced a certain *type* of organism, and (2) its individual development in one or more environments. These two histories determine the range of interactions that the bird can have, and thus its range of knowledge.

I wish to combine now this simple idea, that perception has a biological basis, with two other simple ideas: that intelligence arises out of perception and other biological structures, and that science is a social product of intelligence. Armed with these three simple ideas, I intend to develop a complete epistemological relativism. For the sake of exposition, however, I will discuss the matter as if there were three levels of relativism: perception, intelligence, and science. As we sill see, the arguments used to support relativism at one level, will apply equally well at the other levels. The complete epistemological relativism that results, can be defended form the standard objections to relativism and offers distinct philosophical advantages over sophisticated realism.

2. Perceptual Relativism

Let me return briefly to the bird and its perceptual apparatus. Once we are in a biological frame of mind, the question of why a particular perception is appropriate finds a clear answer: because it succeeds. With the theory of evolution in the background, we should say that the bird has that kind of perceptual apparatus because it served its ancestors well. As long as that success continues, organisms of that type will continue to perceive the world in that manner. It is also clear that different types of organisms will interact differently with the world. And thus, since what serves them well may also be different, different organisms may well have different perceptions of the world.

The foregoing should not lead us to conclude that all perceptions are equal. There are differences in the quality of perceptions. Some clearly allow one organism a better performance vis a vis the environment than that open to a different organism with different perceptions. Eagles, for example, presumably have better eyesight than dogs. Nevertheless, we should not lose sight of a very important point: no matter how successful a perceptual apparatus is, natural selection could have brought about a different one that is equally successful (and it often did). Even an extremely successful perception needs to cohere with the rest of the organism. But very different organisms may establish that coherence in a wide variety of ways. All this is just another way of saying that natural history has many paths open to it. However we put the point, the result is that there *may* be many equally "good" alternative perceptions.

As I have intimated, this result includes those cases in which perception cannot be improved (in the sense that the individual organism perceives as well as it is possible for organisms of that type, while the "design" of the type permits, in that one aspect, perfect adaptation to the relevant environment). For even then, natural history could have chosen different paths in the development of perceptual mechanisms and brains. In our own natural history, for example, it seems fortunate that a big catastrophe befell the dinosaurs, probably the famous Alvarez asteroid. For without such a catastrophe, our mammalian ancestors would not have had the many rich opportunities that the disappearance of the dinosaurs opened to them. Without those opportunities, the growth in size and sophistication required for high mammalian intelligence would not have occurred. Accidents such as that are very common in the history of our planet. And if other planets were to have life on them, we should realize that small planetary differences can lead to completely different histories of life. A small difference in mass, for example, would yield a small difference in gravitation. This alone may be enough to cause a difference in geophysics and in atmospheric density. These two differences in turn would bring about different chemistries and climates. And these last two, of course, would have a great impact upon the development of life on any planet.

As we have seen, what is appropriate for one kind of organism need not be at all appropriate for others. Biology is then bound to take many different paths. But we have also seen that different paths, in this case different perceptual mechanisms, may be equally good. Once more, by this I mean simply that no matter how successful an interaction with the environment is, e.g., a perception, that there could be an alternative interaction which is as successful. And now we arrive at a simple but important point. If two interactions are equally successful, i.e., "good," it is difficult to say that one is superior to the other. Indeed it would be arbitrary to say that one should be preferred to all others.

In previous work, I have referred to the potentialities of a species' genotype as a "frame of reference."[1] I have made the point there, that no matter how good a frame of reference is, there may be others just as good (i.e., that provide for interactions, or performance, that cannot be considered inferior). It would be arbitrary, then, to say of any one frame of reference that the perceptions or views of the world that originate within it correspond to reality, or tell us the way things really are. For it is clear that the others would be just as deserving the honor.

This result would not be affected even if it turns out that one frame of reference (or in this case, perceptual mechanism) is in fact superior to all others. For it would be a mere accident of natural history that no comparable frames of reference (or perceptual mechanisms) have been produced in this planet or in other parts of the universe. One perception, or point of view, can be said to "correspond" to the way things really are, i.e., to be the true representations. At this point, it is useful to remind ourselves of another simple point, one that we take from the special theory of relativity: when a property, e.g., mass or length, can be measured only relative to a frame of reference, and when there is no preferred frame of reference, there is no "naked" instance of that property — this in case, there is no absolute mass or length.

Similarly, if the "way things really are" is perceived only relative to a frame of reference, and since there are no preferred frames, there is no absolute reality either; that is, there is no such thing as "the way things really are," no structure of the universe," no naked truth.

This epistemic relativity would be of little interest if I were to restrict it to perception. I will argue presently that it applies equally well to intelligence and science. But before doing so, I would like to make clear that my analogy to the special theory of relativity is purely heuristic; and that the force of my point does not depend on the truth of such theory; nor is it diminished by the absolutistic character of Einstein's mechanics at a higher level. My point pertains only to what can and cannot be concluded about reality when certain relativistic conditions are met.

3. Intelligence and Scientific Relativism

The reasons for taking epistemic relativity beyond perception can be found in the close connection between intelligence and the complexity of the central nervous system. Two crucial features of intelligence, it seems to me, are its flexibility and its capacity for indirect action.[2] This description of intelligence at the psychological level is buttressed by an analysis of the underlying neural structures. Even at the level of perception we begin to find intelligent characteristics in complex central nervous systems. Whereas a primitive nervous system may process "information" with few degrees of freedom, a complex one can do so in a great variety of ways (see Munévar 1981: 40-44).

Take vision. Some simple organisms detect light just so they can move in the direction of its source or away from it. Less simple organisms can make rather rigid "representations," or respond rigidly to a color or other visual feature. A complex system coordinates many structures, and as a result, it compares sense modalities. It is not far-fetching to say that we wee with our whole bodies. Even though our eyes change position continuously, the image often remains stationary in our visual field. The reason is that there are connections between the visual areas of the brain and the inner ear and the skeletal muscles (as well as the eye muscles), with the result that the brain keeps track of the position of the body, compensating for a variety of motions as it may be appropriate. Vision is also affected by hearing, smell, and the other senses. What may have been an amorphous heap in a dark alley turns instantly into the clear image of a guard dog when we hear its threatening growl. Vision is also affected by memory and imagination, as evidenced by the education of a naturalist that allows him to distinguish insects from leaves, and by the fear of a child lost in a forest. Even behavioral success or failure may influence vision. Experimental subjects first experience a distorted room as if it were square, but when they try to perform in the room, e.g., to touch a mark on the wall, they fail. Repeated failure eventually leads to a gestalt switch, as a result of which the subjects now perceive the room as distorted.

The increase in the complexity of a central nervous system brings about new structures to handle new functions, new connections and coordinating structures, and a general increase in the capacity to route, store, combine, and manipulate information. Here lies the key for understanding intelligence.

How these new functions are performed internally and how the relevant brain structures appear and develop depend on a very long sequence of interactions between succeeding organisms and succeeding environments. In other words, the workings of a central nervous system, or its equivalent, are the product of natural history and of individual development. In this sense the modes of thought depend on evolution. The point, just as in the case of perception, is that natural history could have taken a different path, and thus that there could be other modes of thought that are just as good. As the brains, or their equivalents, evolve through the vicissitudes of different natural histories, they will face different selection problems and pressures, will coordinate different perceptual mechanisms, and will find convenient or economical different structures and solutions. And these are the brains that some day, probably in social cooperation, will invent science.

We can see, then, how the biological underpinnings of intelligence permit us to extend the epistemological relativism from the level of perception to that of intelligence. And we can also see how the biological foundations of science, via intelligence, extend that relativism to science itself. Indeed, in the case of science the relativism may be even more readily apparent; for intelligent beings of different types would construct their science starting form different biological origins (and eventually different modes of thought), and then would experience different social histories. When we discussed intelligence, we realized that the increase in the complexity of the central nervous system offered an increase in the variety of certain kinds of response, while the actual development of that system was clearly the result of a series of evolutionary compromises. In the case of science, social structure may bring about an even greater variety and flexibility of response. In science, thus, the many possible superpositions of social upon natural histories have a multiplicity of paths available to them. Therefore, by the same reasoning employed in the case of perception, no matter how good a "conceptual" frame of reference (i.e., the conceptual potentialities of a genotype) is, there could be others just as good. And as in the case of perception, it would be arbitrary to prefer one to those others.

4. Scientific Convergence?

Nevertheless, it is often thought that advanced sciences should tend to converge. If this were so, science would remedy the relativism that may still be present at the levels of perception and intelligence. The argument goes as follows: even if we grant that natural history may have produced different brains, and thus different modes of thought, intellectual convergence should still be expected with the growth of science. The reason is that science deals, or tries to deal, with all-pervasive features of the universe. These all-pervasive features will pressure different sciences in the same direction: that of the real features of the universe. The more a science advances, the more similar to other advanced sciences it should become. The universe itself will thus pressure the scientific mind as the ocean forces fishes and dolphins to adopt similar shapes. Indeed, in nature we find many examples of convergence: the camera eyes of humans and squids, and the morphology and behavior of marsupials and placental wolves. These are all cases where in spite of different biological starting points, having to deal with the same environmental features led to strong similarities. Our brains may be very different from those that natural history may have produced in another planet, or from those that it would have produced in our own planet had one more or one less accident taken place. But surely, to succeed scientifically they must respond to the all-pervasive features of the universe. Hence, the greater their success, the greater their convergence.[3]

This argument, however, assumes a direct pressure by the universe upon the development of science. And we may recall that one of the principal characteristics of intelligence is precisely the indirect fashion in which it can handle the environment. Thus we should expect that scientific intelligence would be able to handle the pressure of the universe in a great variety of ways; particularly when we consider that in addition to the flexibility of individual intelligence, the social dimension of science provides a second level of flexibility. A genotype that produces a complex central nervous system and that leads to the formation of social groups for the satisfaction of curiosity (i.e., the investigation of the world), is also likely to produce a flexible phenotype. This flexibility in turn often allows for the capacity to adapt to a variety of environments or to a changing environment. The social behavior that produces scientific views, thus, need not be constrained so that those views will converge, even if we could speak sensibly of the pressure inflicted by the same features of the universe. Thermal springs deep in ocean waters kill fish but make many bacteria thrive. What is an opportunity for some organisms is a disaster for others and of no concern to still others. Different brains, or their equivalents, will approach those "same" features from different vantage points. even the same features will not produce the same pressure upon all developing sciences. Since different aspects will be of importance to some brains more than to others, both in guidance and in inspiration, we should, if anything, expect divergence at the level of science. We should not forget, incidentally, that the same ocean that forces similar shapes on dolphins and fishes also produces crabs and squids.

In any event, the defense of relativism requires only that convergence need not be the case. For then natural history could have brought about alternative ways not only of perception but of conceptualization as well.[4] Conceptual relativity in understanding the universe, the absence of an absolute frame of reference, destroys the claim that scientific views correspond to reality, that they will ever, or that they may approximate such correspondence (approximation requires convergence). We have no warrant to make such a claim because it would be arbitrary to say that the views developed within any one frame of reference correspond to the way things really are — since there may be other frames of reference that are equally good. This situation leads to the relativization of the notion of reality as well, for the reasons given earlier in the analogy to the special theory of relativity. This completes my positive presentation of evolutionary relativism.

III. Objections To Relativism

1. Standard Objections

Let us see now how this evolutionary relativism can withstand the standard objections against traditional relativism.

    1. Relativism is contradictory. This is an objection against simpleminded relativism, according to which truth is relative to each observer. Presumably this would permit one observer to hold that A is true while another holds that *not A* is true. Why this situation would lead to a contradiction is not clear, since truth in this case is relative. At any rate, in evolutionary relativism we do not hold that A and *not A* are true together, for A is held in one frame of reference and *not A* in another.

    2. Plato's first argument (in the Theætetus). According to Plato, Progagoras held that all points of view were equally valid. If that were so, the absolutist point of view would be valid as well. But the absolutist point of view claims that relativism is wrong. Therefore, Plato concludes, relativism is incoherent. I am not sure of the strength of this argument; but in any event, it does not apply to evolutionary relativism, since I do not claim that all frames of reference are equally valid, but only that some may be.

    3. Relativism entails that the universe would not exist without observers. If the way the world is, reality, is relativized to a conceptual frame of reference, we are tying the existence of the world to a frame of reference. But surely, as Tuomela, arguing in favor of realism, puts it, " ... at least some real objects may exist in the world even if ... all mankind is destroyed and even if there had never been any human beings" (Tuomela 1985: 109). Objections of this sort are sometimes discussed by physicists concerned with the seemingly relativistic implications of quantum theory. The universe must be independent of any frames of reference. If it were not, then it would not come into existence until it could be described within the point of view of some observer or other (presumably the sort of thing that I mean when I talk of describing the universe within a frame of reference), and it would go out of existence when there are no longer any observers to describe it. Clearly, this consequence reduces relativism to absurdity. But this objection misses the point of evolutionary relativism. The frames of reference in question need not be actual frames. Relativism requires only potential frames of reference, therefore the objection does not apply. I would like to explain the point by means of an analogy to the special theory of relativity once more. The equivalent objection, with respect to absolute, real mass, would run something like this: you say that mass is relative to a frame of reference; are we to conclude that if no one were around to measure mass, objects would have no mass? Of course that would not be so. The frames of reference, or measurement, in question may all be potential.

4. *(This objection is considered decisive by many philosophers.)* Evolutionary relativism itself is expressed within a frame of reference, is it not? But why should that frame of reference be preferred to all others? To put the point another way, is the thesis of evolutionary relativism absolutely true or not? If it is, my position turns out to be absolutist at the meta-level. It it is not, it will then be either false or else "true" only in some subjective or relativistic way. In this case why should anyone agree with me, or even take evolutionary relativism seriously? It seems to me, however, that this objection begs the question. For it supposes that only theses which are candidates to absolute truths, or at least to approximations to such, can be taken seriously, let alone deserve agreement. And this presupposition appears in a context of discussion in which the notion of absolute truth is precisely what is in question. An alternative to a standard is dismissed because it does not conform to that standard.

2. *Sophisticated Realism*

I would like to go on now to more interesting objections that come not from pot-shots taken at relativism but are rather born from the development or sophisticated forms of realism. I do not include here so-called internal realisms because I think that they are just badly disguised forms of relativism. According to Tuomela, for example, it is characteristic of internal realism to hold that:

> (a) the question, "What objects does the world consist of?" only makes sense within a theory or description. Accordingly, the world is in a sense "man-made" or "processed" through a human conceptual scheme ... (b) the world can be described in several true and complete but in some sense rival ways. Furthermore ... (c) truth is an epistemic (and theory-dependent) notion. (Tuomela 1985: 96)

None of these theses are inconsistent with evolutionary relativism; and indeed the task of trying to make realism from such an obviously relativistic view is by no means easy. Tuomela, for example, has to juggle claims as seemingly contradictory as "There are real particulars (objects, events, processes, etc.) which are mind-independent," and "There is no ontologically given, categorically ready-made world."[5] Of the arguments he provides for his realism, the strongest one that may serve against relativism has already been mentioned; it amounts to the discovery that if the world were relative to a conceptual scheme (or, in my account, to a frame of reference) then it would not exist without us, which presumably is absurd. But we have seen that this argument naively takes the frames of reference in question to be actual frames, whereas the theory of evolutionary relativism only requires potential frames.

I will not pursue this matter further here because it does not serve the purpose of this paper to discuss in detail the failings of views that are already close to relativism. It is more pressing to see how a sophisticated realism can try to take stock of the biological considerations advanced by evolutionary relativism. Let us take a realist who might grant that natural history could bring about, or could have brought about, alternative but equally good frames of reference (e.g., Clifford Hooker (1987), or perhaps C.I. Lewis (1929, Chap. 6)). In Hooker's view, all those frames would result from interactions with the universe. It is the function of the scientist, or rather the philosopher-scientist, to search for invariants between those frames or reference, just as it is the scientist's job generally to look for invariants in nature. The goal of scientific or naturalistic philosophy would be to obtain a theory of those cognitive invariants. That theory, of course, would be a true theory of nature. A scientific, naturalistic realism may still triumph in the end.

I have two comments and one reply to make. The first comment is that a complete theory of invariants may be in principle unattainable in an enterprise that would be to a large extent historical. The second is that the resulting realism — if the theory of invariants is obtained, that is — would be very strange. For reality would not be described at the customary level of physics, chemistry, and so on, but at that of cognitive invariants. Although perhaps this unique description of reality would legitimize the relativization of science at the more customary level, my reply is very simple: that super-theory of invariants must be developed within a frame of reference. But then there could be alternative theories of cognitive invariants developed within alternative frames of reference. And, again, some of those frames may be equally good, and so on. Relativity is established at the meta-level as well (and at any meta-meta-levels that may pop up). This result would not vary if perchance scientists from species with different but equally good frames of reference agreed on a theory of cognitive invariants. For again, it would merely be an accident of natural history that only such lines of cognition had been developed.

A similar reply can be made to C.I. Lewis, who would equate reality with the conjunction or collection of all the best points of view (or, in my system, of all the best frames of reference). If the interaction between the universe and different modes of cognition can produce a variety of representations of the world, once we have the collection of representations we should have an exhaustive account of what the world is like. This reasoning by Lewis is at first sight very attractive, but Lewis assumes that such an account can be given independently of a point of view (we would discover what cognitions of the world each of the alternative frames can produce, and then we would collect all those internally produced cognitions). It seems to me, however, that there is no such neutral or preferred frame of reference in which to carry out this investigation, and thus no neutral or preferred description.

3. Confusion

Some realists often become confused at this juncture of the argument. The objections seem to have been answered. But have I not assumed the truth of the theory of evolution, of certain parts of neuroscience, and of other ideas? Does not my view assume realism, then? I will answer in two stages. First, as an argument against realism, assuming the truth of some important contemporary scientific views is unobjectionable. As in any *reductio ad absurdum*, we take for granted premises that are dear to the other side, and then we derive conclusions unacceptable to that other side. I confess, however, that I wish to defend relativism as a better way of understanding science. To that effect I would grant that I accept, say, the theory of evolutionary biology as true, but true only in a relative sense. My guess is that evolution is true for us, or at least for some of us. At the same time, I am willing to grant that evolutionary biology may well not be true, and perhaps not make any sense at all, within other frames of reference equal or superior to our own. I do not find anything perplexing in this answer; and I suspect that, if the realist does, it is only because he cannot imagine what a relative truth might be. Of course, in this century many philosophers have spoken of truth relative to a language or to a conceptual scheme. And we can play at seeing how a sentence is true in some special ("internal") sense if its utterance conforms to certain rules for making statements within a point of view. And so it should suffice to say that such and such a point of view, e.g., the theory of evolution, is true, or on the way to being true, relative to our frame of reference. But realists — at least some realists — no longer want to play when we go beyond the truth of sentences (normally the uninteresting sentences given in examples taken from philosophy of language) to the truth of whole points of view such as the theory of evolution. They want to know how truth — real honest-to-goodness truth — can be relative. I will tell them.

IV. Relative Truth

1. Development

In the hypothetical comparisons between frames of reference, when two frames led to similar performance, it was found arbitrary to make of either one a preferred or absolute frame. This indicates that the notion of performance can be fruitfully tied to the notion of understanding, particularly to that of scientific understanding.[6] In the rest of this paper I will use this indication to suggest a biological theory of relative truth. I will introduce that theory by means of an illustration. Let me suppose that when I perceive a particular apple I see it as red, taste it as juicy and delicious, and find it beautiful enough to make it the subject of a still-life painting. These and other perceptions of the apple, let me suppose further, are the ones that serve me best to deal with that part of the world (the apple). Imagine now that beings of a very different kind have perceptions of the apple which are very different from mine, though just as good as mine. Upon coming to know of these beings' perceptions, should I stop trusting my perception of the apple? Should I replace it by the perceptions those beings have? I suspect the answer in both cases is "no." For I have already said that this is the best way I can perceive that part of the world. Thus knowing of the other beings' perceptions would only lead me to conclude that I do not perceive "the way the apple really is." For I realize that to do so would be arbitrary. But in this instance I do not need to change my perceptions of the apple.

In this example, my perceptions best exploit the resources of my genotype, (or rather, of the genotype of beings like me), in dealing with a typical environment. Were the environment to change, those perceptions may no longer be so appropriate. And, of course, in our mundane dealings with the world many perceptions, if not most of them, seldom allow us to reach that level of performance. But whenever the resulting performance is as satisfactory as in the case of the ideal perceptions of the apple, we tend to think that the world must be as we perceive it. It is then that we feel entitled to speak of true representations.

I wish to suggest a theory of truth along these lines. Of course, in our conceptualizations of the world — in our scientific views — we seldom, if ever, reach the level of sufficiency or satisfaction given in the case of the apple. Nevertheless, when we approach it we speak of truth. Human science is ultimately a variety of human behavior, and I consider human behavior part of the human phenotype. Although, at least in the case of humans, we should speak of phenotypes, since the plasticity of human behavior is such that there could be many expressions of the genotype even in the same environment. It seems to me also, that some phenotypic expressions exploit better the resources of the genotype in a given environment. Likewise, some scientific viewpoints (with their complex machinery of practices, experimental procedures, and so on) permit us to exploit better the resources of our genotype in a given environment (e.g., in dealing with the dynamics of bodies). In other words, some viewpoints enable us to realize more of our potential for performance. In this biological context, a viewpoint is said to be relatively true when it approaches the limits of the resources of the genotype. When a theory allows us to deal with the world in a great variety of ways, when thinking that the world is as the theory tells us leads to continuing success, when this capacity for performance clearly surpasses that of its competitors, then we come to think that the world must be so. And in a limited domain we may not be able to conceptualize the world any better. We conceive the world as powerfully as in the earlier example we perceived the apple. It is then that we speak of truth.

2. Comparison with other Theories of Truth

It may be felt that this account merely explains why we feel the impulse to say that some "representations" of the world are true. But I have done more than to provide that biological basis for a psychological account of truth. For this account explains also why it may be worthwhile to make distinctions between true and untrue. Normally we would say that a viewpoint is true because the interaction (with the world) that results is of great quality (or seems to be) and very superior to its alternatives. This is not to say that we have finally arrived at the way things really are, but merely that our "picturing" of the world approaches the level of quality exemplified earlier by our perception of the apple. This "picturing," just as that perception, is nonetheless relative to a frame of reference, and thus the truth involved is relative truth. Many realists have tried to explain the success of science in terms of truth. A viewpoint was successful either because it was true or because it approached truth. My account of relative truth turns the tables. The relative truth (or seeming absolute truth) of a viewpoint depends on its success, not the other way around.

This point bears expanding. Some philosophers may argue that I have only provided a descriptive or psychological account of truth. But, of course, what else could a naturalistic account be? The naturalist's task is to explain why a "picture"-making activity appears satisfactory. When we see the world we see it as something or other. Some of this "seeing as" holds us in a strong grip because it permits such a strong interaction with the world (the picture-making is after all only part of a means of interacting with the world, as we saw earlier in the case of perception; cf. my remarks on the complexity of the central nervous system and the flexibility of perception). I suggest that it is that grip under those conditions of successful interaction that seems to have a special character. I believe that character is what philosophers have been trying to explain with correspondence theories of truth.

Some qualifications are in order. Few views are so successful that they are accepted on the basis of a clearly superior record. They are accepted because in a few instances the success achieved is felt to be so striking that many members of the discipline find that way of doing things extremely promising. That is, they are accepted on the basis of promise of performance rather than on total performance. After a group takes up a way of thinking about the world, and elaborates it to the point that its performance begins to approach the limit of the potential of the genotype in the relevant environments, then its truth seems evident to all those concerned. There are also cases in which that limit is not approached but the scientists committed to the point of view are unable to think about the world in any other way, and so they keep on feeling that the truth must lie somewhere along the path they have undertaken.

There are, in addition, cases in which a point of view, if developed, would have exploited better the resources of the genotype; and so years later we feel that an opportunity has been missed. And I suppose there are cases in which the superiority of a point of view goes unrecognized. All the sensible things that philosophers wanted to convey with the old correspondence notions can be conveyed with this relativistic notion.

As for the objection that I must accept the truth of evolutionary theory, perhaps it is clear now why my answer is "yes": I accept its relative truth. That is, I believe that the evolutionary thinking is the best I can do within the bounds of my conceptual equipment, and I suspect that it has the highest potential for performance with respect to a great number of areas of experience, particularly those that have to do with living things and their history. My belief may turn out to be wrong, but right now I am only trying to explain what I mean when I say that I believe that the evolutionary point of view is true (or that I am committed to its truth or words to that effect). To alien beings elsewhere, engaged in completely different modes of interaction with the universe, evolutionary thinking of any kind resembling ours may not make any sense within the bounds of their conceptual equipment. But to beings like us it does. Or so I believe. I would say similar things about the truth of my philosophical position, if called upon to do so; and I would adduce as evidence precisely the evolutionary arguments I have provided all along in this paper.

Nothing in this account of evolutionary relativism should be taken to suggest that every scientific idea must somehow be tied directly to our dealing with the world, that it must have practical consequences of some sort. I have already mentioned the indirect action of scientific intelligence, and I would like to add that much, if not most, scientific work is devoted to the articulation (conceptual, mathematical, experimental, technological) of our views about the world (see Munévar 1981, Chap . 4). It is in this feature of science that we can find use for some of the more sensible intuitions from a coherence theory of truth.

When philosophers noticed that not every scientific idea could be "compared" with the world, it was easy to realize that some ideas are accepted by how well they fit in with other ideas that have already been accepted. (Not that any such idea can be so compared — what they mean is that the ideas can be tested for the purpose of confirmation or refutation.) The hope arose occasionally, then, that coherence could replace the beleaguered correspondence as criterion of empirical truth (the influence of logic has played a notable role, or perhaps I should say a notorious one, in the development of this hope; but I will not touch on that subject here). I think it would be a mistake to demand that increasing coherence (if it can be determined) must be accepted as evidence of truth. Science is not very old, and it is therefore sensible to suppose that we do not quite have a perfect grasp on nature. Given that supposition, we may also imagine that sometimes when new ideas cohere with entrenched ones, we are faced not with progress, but with stagnation; for there are cases in which daring in thinking is called for to uproot outmoded forms of though, and thus coherence may only reveal lack of nerve and imagination. Nevertheless, if articulation is the main purpose of much of science, ideas or points of view that permit certain articulations increase also our potential for performance. In this respect, an increase in what may seem like coherence would fit in well with the notion of truth of evolutionary relativism.

The notion of ideal truth implicit in evolutionary relativism has some features in common with a pragmatist's ideal truth. But only to a point. Apart from the biological context in which we can now understand the approaching of a limit, it is important to realize that in the evolutionary account the limit may well be a horizon that recedes. There are several reasons for this. The exploitation of the resources of the genotype depends on the environment, or environments, involved in the interaction. The environment is not static, however: it may simply change, in which case the phenotypes that may have been quite adequate before may be challenged successfully by a different approach more in consonance with the new circumstances. The relevant environment may also change as the result of the interaction with the first successful phenotype. That is, the environment may be transformed. But it may also happen that success in one environment leads the organisms in question to venture into other environments, which jointly offer challenges of a different sort. And finally, the influx of new ideas, as well as the articulation of the main points of view, change the conditions of interaction with the universe; and that change of conditions opens the door to a different approach.

These considerations make the relativistic ideal of truth a changeable one. Perhaps we can now understand why scientists so often feel that they are pretty much in possession of the truth, while at the same time allowing that others who had similar feelings in the past were "wrong," and even that others in the future may feel just as strongly about different ideas. At times, we may come close to truth with respect to a certain environment, just as some species may come close to equilibrium with a given environment: but as conditions change, other ideas will be more appropriate and thus considered true.

3. Social Relativism

The flexibility of the human phenotype, particularly at the level of social forms, raises the possibility that alternative social histories may produce alternative views of nature that are equally good. If to that result we add the possibility of the transformation of the environment and of the expansion into different areas of experience, to say nothing of the actual shortcomings of any humanly constructed points of view, social relativism appears no longer as a problem but rather as a reasonable epistemic strategy. Relativism within a frame of reference (intra-species relativism) may thus be established within the more general position that establishes relativism between frames of reference. There are not only different ways of approaching the limits of the resources of a genotype, but even different limits to approach, according to how the future natural history is to be affected by the choices made along the way. And if social relativism in matters of science can be a fruitful way to proceed, it would seem that in other matters we should look at it with something less than philosophical revulsion.

V. Conclusion

This account of relative truth shows, I believe, how evolutionary relativism meets Plato's second objection to Protagoras; namely that if truth were relative to a culture, or to a point of view, then there would be neither reason nor motive for changing (for every point of view would already be satisfactory).[7] This removes one presumed advantage of realism over relativism. In other respects the balance of advantages and disadvantages has now tilted in relativism's favor.

Consider, for example, the ability to speak about reality, presumably one of realism's dearest projects. The sophisticated causal realism of Hooker, Lewis, and others cannot tell us what the world is like anymore. We do not know it, we cannot know it. The world is that "something" that in causal interaction with frames of reference brings about certain points of view. It is the same world, but forever indescribable: a "mysterious substratum," a "kingdom of Being," "noumena." The scientific realists end up thinking what cannot be said, and transcending what can be thought. Or at least trying to. The evolutionary relativist, on the other hand, has no such problem. The world is as his science tells him.[8] He may successfully conceive the world a certain way, just as he may successfully perceive the apple a certain way. Of course, at a different time, or within a different frame of reference, the world may be quite different. Truth is, after all, relative.

I would like to end with an invitation. I have heard some realists comment that there is not much difference between evolutionary relativism and the sort of sophisticated realism they favor. I agree. I would call on them to realize that except for the emotional connotations of the title, evolutionary relativism is the view that best fits their philosophical outlook. Let us outgrow adolescent fantasies about the naked truth and work together on the task of understanding the nature of knowledge.

---

* I thank the many people who have commented upon the preprints of the paper. I wish to express my gratitude to my good friends Clifford Hooker and Andrés Rivadulla for influencing the final shape that the paper took, mostly by disagreeing vehemently with my views. I wish I could report that they have acquired the wisdom to see things my way.

Notes

1. See Munévar (1981), particularly Chap.3.
2. As Piaget pointed out, thanks to intelligence an organism's scope of interaction with the world goes beyond immediate and momentary contacts (Piaget 1950)
3. This argument conveys a generalized opinion in the scientific fields that attempt to detect signals from advanced alien civilizations, but it is older than those fields. An earlier version of it may be found in Lorenz (1971: 290)
4. Sometimes it is argued that it is pointless to talk about alternatives whose contents we cannot discern. According to this position, talk of the possibility of alternatives should not be permitted unless the alternatives themselves are specified. And according to a well-known argument by D. Davidson and B. Stroud, if the alternatives can be specified they are not really alternatives. I find this attitude unreasonable. The first motivation for it comes from a misguided constructivism. It is misguided because that same attitude has been on the losing side often enough in the history of science, and because in evolutionary relativism we do not just raise idle possibilities but rather provide an argument from biology to the effect that different modes of thought would be the result of different natural histories. We may not know what will be produced (or "constructed") but we do know that our modes of thought depend on certain conditions: since those conditions do not obtain, we may conclude that the results will be different. The second motivation, the Davidson-Stroud argument, is based principally on the mistaken belief that understanding an alternative requires translation. Although this second motivation is not truly relevant to the discussion of this paper, a detailed look at it may be of interest. See Munévar (1981, Chap. 7).
5. Tuomela (1985: 106). Tuomela tries (pp. 112-114) to solve this problem by means of a double-existential quantification and a hint at an evolutionary account of knowledge. The first quantification is over the possible ways of slicing up the world, and the second over those ways favored in the "best explaining theory." This move simply expresses the problem in logical jargon, but it does not provide a solution. As we will see below, the determination of the possible ways of slicing up the world must be undertaken within a frame of reference. As for Tuomela's own scheme, it is difficult to see what can be accomplished by the first quantification, since it is disregarded and only the second one counts. His talk of vague or indeterminate particulars that science makes determinate neither helps him nor is consistent with the notion that science determines ontology (since we would have two ontological levels, then, as he seems to recognize, pp.113). Furthermore, as we have seen, an evolutionary account of knowledge, his second but vague hope, can take us in very different directions from those that lead to a justification of realism.
6. This task is carried out in detail in Munévar (1981, Chap. 4).
7. The problem of rationality is divorced from that of theory choice in my evolutionary epistemology. In several publications I have developed a social conception of scientific rationality (e.g. 1981, Chap. 4, and especially Chs. 9 and 10, this volume).
8. To be more precise, as his "successful" science tells him. "Successful" is used here in terms of performance, as it has been throughout the paper.

References

Feyerabend, P.K. (1975). *Against Method*. London: NLB.
Hooker, C. (1987). *A Realistic Theory of Science*. Albany: State University of New York Press.
Kuhn, T.S. (1970). *The Structure of Scientific Revolutions*, 2nd ed. Chicago: The University of Chicago Press.
Lewis, C.I. (1929). *Mind and the World Order*. New York: Dover.
Lorenz, K. (1971). *Studies in Animal Behavior*. Cambridge, Mass.: Harvard University Press.
Munévar, G. (1981). *Radical Knowledge: A Philosophical Inquiry into the Nature and Limits of Science*. Indianapolis: Hackett.
Piaget, J. (1950). *The Psychology of Intelligence*, transl. M. Piercy and D.E. Berlyn. London: Routledge and Kegan Paul.
Tuomela, R. (1985). *Science, Action and Reality*. Dordrecht: Reidel.

2 Extraterrestrial and Human Science

Travel teaches us not only about other places and people but about ourselves. Likewise trying to understand what other intelligent life might be like teaches us about our own intelligence. And trying to understand how alien intelligence may view nature teaches us about what our own views of nature amount to.[1] In the scientific program to search for extraterrestrial intelligence (SETI) we find almost bare many common assumptions about the origin, development, and nature of science. Thus from an analysis of SETI we may be able to draw some interesting philosophical lessons.

In this paper I will be concerned mainly with three notions that are frequently advanced by SETI proponents. The first notion is that once life appears on a planet, intelligent life is also very likely. The second is that once intelligence appears on a planet, science itself is likely. The third is that all scientific civilizations have something in common (i.e., the overlap in their scientific views of the world) and thus the basis for the beginning of communication between them.[2] The first two notions are advanced to support the contention that there is probably someone to look for. The third to indicate that contact will not go for naught. But in spite of their initial plausibility, I will argue that these notions are plagued with less than obvious assumptions at many levels, and that they lead to a questionable account of our views of nature.

SETI proponents believe that life can begin elsewhere, that once it begins it is likely to become more complex, and that complexity produces intelligence. Presumably, as time goes on, intelligence will improve its attempts to understand what the world is like; thus begins the almost inevitable road to a technological civilization. Whether life can begin elsewhere is a matter of great controversy. But I will grant for the sake of argument that it could. In the same spirit I will grant that, at least for some time, the complexity of life may increase; and I will also grant that intelligence is the result of certain complex biological organizations. But granting all these crucial assumptions of SETI is not the same as granting that alien technological civilizations are very likely.

Let me take stock of what I have granted. After some primitive form of life appears on a planet it will not remain uniform for long. Small variations in the environment and other factors will bring about diversity.

Of course, diversity is not the same as complexity, but it gets us on the road to it. For diversity means that there will be different kinds of biological structures and different ways of interacting with the environment. And the possibility then arises that eventually some of these structures and functions will combine. A and B may come to work together and a new structure C will arise to coordinate their work. And now A, B, and C together will form a new whole that is more complex than either A or B as separate individuals. During the first couple of billion years of life on Earth, prokaryote cells were the most prevalent, perhaps the only form of life. These cells in which the chromosomes are not protected inside a membrane (the nucleus) eventually led to cells with nuclei (eukaryotes), which are more complex. According to Lynn Margulis this important step came about by the symbiosis of different kinds of prokaryote cells.[3] In any event, once cells with nuclei appeared it was possible to form organisms that combine many of these cells, sometimes billions of them. These organisms are of course very complex wholes of eukaryote cells that perform many different but coordinated functions. Although after billions of years the increase in the complexity of life can be considerable, complexity is not always bound to increase with time. Changes in the environment of a planet, some of them caused by life itself, may make it very difficult for all but simple organisms to survive on that planet.

Let me concentrate now on a particularly interesting kind of complexity. Eventually some Earth animals developed intricate patterns of muscles and bones so they could move about, external senses to give them information about the world, and internal senses to monitor a variety of organs. It does not take much to see the advantage of coordinating these functions. A successful predator not only sees the prey but can move so as to catch it. On Earth, a popular answer to this problem of coordination is the central nervous system. And this is an interesting answer because it is in connection with a highly complex central nervous system that intelligence becomes conspicuous. A highly complex central nervous system is not limited to just one way of handling the information that it receives from the world: it can rout and combine information in a variety of ways, it can compare sense modalities, it can store information and consider alternative actions, that is, it can make use of memory and imagination. Visual perception is a good illustration of this kind of complexity. At a very primitive level we may suppose that the detection of light is enough for a certain organism, in order to move towards or avoid the light. The next step comes when the organism gains an advantage by being able to discriminate visually between objects, which may be achieved by making internal representations of those objects. These representations grow in sophistication and the corresponding nervous structures in complexity, when the "input" from the eyes is coordinated with that from other senses. For example, when we are looking at a painting of a group of people our eyes are not stationary. First of all, the eye muscles make the eyes scan continuously. Second, our heads move sideways as well as up and down. Our whole bodies may also move, carrying our heads, and thus our eyes along. But the images of those people remain stationary. This is very different from, say, a video camera, whose images do move up and down or sideways, the more so the more unskilled we are at shooting with it. The reason our perceived images remain stationary is that the brain takes into account the automatic movement of the eyes as well as our body position in order to arrive at a perception that we can handle. The brain takes into account our body position by receiving information from the inner ear, which keeps track of the inclination of the body with respect to the Earth's gravitational lines of force, and from hundreds of skeletal muscles. But visual perception is also easily affected by the other senses. Take hearing. As we walk down a dark street at night we may perceive some bundles a few steps ahead. But at least one of those bundles suddenly becomes a sharp image when we hear the distinct growling of a guard dog. Perception also takes into account memory and imagination. An artist well trained in the history of art may see many more details in the painting than most of us can, just as a well trained naturalist can detect a rare bird in a bush where most of us could see only foliage.

The more complex the central nervous system, the more complex the relationship between the organism and the environment, for the organism gains more degrees of freedom. Thus intelligence arises out of perception and other biological structures as the complexity of those structures increases. This account agrees with Piaget's description of intelligence as an instrument of adaptation not necessarily tied to the immediate and momentary demands of the environment (human beings, for example, can figure out solutions to problems that will confront them far away and years hence).

Let me sketch now the main hurdles that life has to overcome on its way to an advanced technological civilization. On this account, to say that intelligence is adaptive is to say that a highly complex central nervous system (or its equivalent) is adaptive. But then intelligence is adaptive only for certain kinds of organisms and not for others. It would be adaptive for primates, for example, but not for cockroaches. Let me illustrate the point by means of an analogy. It is well known that the opposable thumb is a highly adaptive feature of human beings. But it would not be so for horses. And it does not even make any sense to ask whether it would be for cockroaches, since roaches do not have the kinds of physical structures to which opposable thumbs can be attached. We might think that roaches would be better off nevertheless if they were smarter. But to put the point properly we have to consider whether roaches would be better off with a more complex brain. And now we may begin to see the difficulty: there is a price to pay all along the way to intelligence. The price is that a complex brain demands a high metabolism. In a minor way the same point may be made about sight, which also seems to be quite an advantage. Imagine that a population of small mammals has come to live in dark caves. The structures that permit sight use a lot of energy, and so these mammals have to spend much time and work getting that energy. Since sight is now of marginal advantage, the mammals that preserve sight are not as competitive as others that use only a fraction of that energy to enjoy improved hearing, touch and smell. It would be nice to have sight, but a mammal of that size can't afford the price to keep it. And for a population of sightless mammals it would make no sense to develop it. Likewise, an increase in the complexity of the brain requires that the organisms of the species in question gain some advantages that compensate for the price in metabolism that they have to pay. In the case of many species on Earth, including ours, those advantages have been there. But we should not expect that they would be there on any other planet where life may evolve.

Consider our kind of intelligence: mammalian intelligence. If the dinosaurs had not become extinct, mammals would have remained small vermin. Large mammals could not evolve because an increase in size would make it easier for dinosaurs to prey on them. But the price that mammals would have to pay for a bigger and more complex brain would probably be a bigger body. The point is that a species, or some other taxonomical category, can be successful enough on a planet to preclude the adaptation by other species that could some day evolve into creatures of high intelligence. In our own day we ourselves are a cap on the development of intelligence by others. Suppose that raccoons became increasingly intelligent. They would become such pests that humans would probably hunt them to extinction. Our very way of life tends to wipe out animals that enter into close competition with us.

Let us imagine a planet very similar to our own. Let us suppose that in that planet also the conquest of the land by fishes would have provided the necessary opportunities for an increase in the complexity of the brain. But let us also suppose that on the planet insects have already appeared on the land and were even more successful than on Earth. Because of their physical constitution, insects are not likely to grow large enough to develop the sort of large brain associated with intelligence. But insects have many adaptive features that serve them quite well. Thus they can be successful without being smart. In that planet they rule the land: any fish that crawls out of the water will be eaten by insects, and if perchance eggs from that species are not only laid but hatched, the young fishes will be devoured. Intelligence as we know it is not likely to arise. The smartest being on that planet would be some kind of octopus. (It does no good to point to whales and dolphins — those are mammals and would have never evolved if vertebrates had not developed on land to begin with.)

Thus on other planets the cap may come from many different kinds of beings, even if their own intelligence is rather modest by our standards. All it takes is that in some other respects they can adapt first to the land, or whatever key environment we consider. But what enters into that timing? Most often just accidents of natural history. For example, it is possible that the disappearance of the dinosaurs may be traced in large measure to the collision of a gigantic asteroid with the Earth. But there is no guarantee, let alone a law of nature, that accidents of natural history are going to favor the development of high intelligence.

Let me imagine, nonetheless, that on some planets central nervous systems as complex as ours, or more complex, do evolve. Will technological civilization then come about? Not automatically. It has to be the right kind of intelligence: technological intelligence. The evolution of human intelligence is tied to the use of tools for hunting and many other purposes. But the evolution of this mode of interaction with the world makes sense only if you have the right kind of body. Dolphins, for example, which are creatures with complex brains and perhaps high intelligence (even if not in a class with ours), have no hands, to say nothing of opposable thumbs. There is a clear sense in which we express our intelligence by having the appropriate bodily interaction with the environment. A "technological" intelligence would not be adaptive unless the right kind of body developed along with it. Spears may have been a sensible option for our ancestors, but harpoons would not have been so sensible an option for the ancestors of dolphins.

Nevertheless let me suppose that technological intelligence does arise and takes over a planet. Technological civilization still does not follow automatically. A technological civilization is in part the result of complex social processes, thus the required type of intelligence must be not only technological but social. But even if we have the evolution of this kind of intelligence, a highly advanced technological civilization may not arise. One reason is that high technology may well require the development of science. On our own planet a turning point came when the new science, culminating in Newton, was able to bring together astronomy and physics. But in a planet very similar to ours but perennially covered by clouds (or in a solar system travelling through a dust cloud) a comparable development of astronomy would be most unlikely.

Imagine, though, that we have a favorable physical environment where intelligent beings (both socially and technologically) can receive the inspiration and rewards that would take them on the scientific paths loaded with the right kinds of intellectual breaks. We still cannot expect an advanced technological civilization. For having the right physical environment is not enough. Social factors may still prevent the development of science as we know it (let alone a more advanced science). It is plausible to suppose that the progress of science requires that ideas may be criticized and that alternative conceptions of the world be developed and defended even if a majority in a group do not agree with them. But in a species biologically inclined to a degree of social cohesion greater than ours, the criticism of the metaphysics of the society (e.g., of their account of the origin and nature of the world) may be seen, or felt, as a threat to the cohesion of the society and put down at once. It seems that if in our world science barely made it, on that other planet science would have no chance.

I do not wish to argue that a technological civilization could not arise on a different planet. My intent is merely to point out that the process is by no means automatic, that it requires many good breaks from natural history.[4] A critic may argue that natural selection could have gotten around most, if not all, of the obstacles I have mentioned. All it takes is a bit of imagination, and we know how imaginative natural selection can be. For example, one of the reasons why advanced technology seems to need a social milieu to exist is that no one human being can fully develop a theory as comprehensive as, say, Newton's mechanics (it took centuries) to say nothing of all the other branches of physics, chemistry, and so on. But even within one school it is difficult enough to come up with a few good ideas. To be able to see their flaws, possible means of improvement, or their connections to other areas of science often requires that we look at those ideas from many different points of view. One human being could not do all this. Science and advanced technology require a division of labor.

Imagine, however, a planet on which a single organism — not a single species, a single organism — comes to dominate even more than human beings do on Earth. This would be a strange organism that covers the environment like a comforter and grows larger by creating more branches on itself until it has finally covered much of the planet (if the food supply decreases this intelligent organism will either "farm" differently or drop off a few branches). Instead of a central brain this organism has something that rather resembles a network of ganglia (large, complex ganglia to be sure). Although the action of the ganglia tends to be coordinated, in a network that large there must also exist a fair degree of decentralization. In that case ideas may be brought up by one particular ganglion and criticized by other ganglia, and so on. The concept of self of this organism may be quite different from ours, but the point is that in a single organism we may find the equivalent of a whole species. So this organism could develop an advanced technology even though in principle it is not social.

It is clear, then, that if certain avenues of development are closed to life, natural selection may find others. But the price of alternative natural histories would be alternative forms of intelligence and eventually alternative ways of formulating views of the world. The reason is that the brains (or their equivalent) that would result from such radically different natural histories would arise from entirely different biological structures, and thus in coordinating these structures the developing brain would face different evolutionary problems and would have different solutions and opportunities at hand. In the neurological ward of a hospital we find that people whose brain structures have been altered have peculiar ways of perceiving and conceiving of the world. Of course, their modes of thought are maladaptive, just as skeletal structures that deviate from the norm may be maladaptive in a human. But for different creatures different brain structures and their corresponding modes of thought may be as adaptive as their different skeletal structures are. The consequence of this point is that the science of a species, or kind of organism, may be relative to its natural and social history. Thus species with very different natural histories may have little overlap in their scientific views of the world. If this is so, there would be much less in common to serve as the basis for inter-stellar communication with other technological civilizations than the proponents of SETI make it out to be.

Defenders of the SETI program often assume that advanced sciences and technologies must exhibit a high degree of convergence. The grounds for this assumption are presumably, that as science grows in scope, the brains that produce that science must reckon with all-pervasive features of the universe. Just as dolphins and fishes have very different evolutionary histories but similar shapes because they both live in water, so sciences that deal successfully with the basic forces of the universe must come to similar views. Nature already offers many cases of convergence: Placental and marsupial wolves, and camera eyes in squids and mammals[5], to mention only two of the most striking. Furthermore, when it comes to communication with advanced technological civilizations, we are talking about species that at a bare minimum have built means of electromagnetic transmission and may also have embarked in a program of space exploration. Their views may be superior to ours (having been around longer) but surely they must overlap with ours to some extent, for at least to some extent they and we are successfully applying the laws of electromagnetism.

Nevertheless the matter is not this straightforward. We must realize that even all-pervasive features of the universe would be interpreted differently by different scientific intelligences. As we have seen, a highly complex brain can deal with the environment in a very flexible and indirect manner. Moreover, it is not one brain but an ensemble of brains in very complex social relations that deal with the universe through science and technology. Whereas in the case of the ocean we had direct pressure (selection) on aquatic animals, in the case of the deep forces of nature we have many different ways of handling the pressure (indeed, a double tier of evolutionary slack). Even in the case of the ocean, animals with very different evolutionary histories have different shapes, as we can tell just by looking at crabs and salmon (fishes and dolphins are much more closely related). The very same "feature" of an environment impinges very differently on different organisms. A hot spring may kill some fish while making bacteria thrive. It is a mistake, therefore, to describe the situation as if different brains were dealing with the same problems. We rather have different brains dealing with different problems. Indeed those different brains will have (1) different starting points for inspiration, (2) different motivations, and (3) different social means of dealing with conceptual matters.

As for the overlap on electromagnetic theory, we should guard against confusing an overlap in performance with an overlap in content. For in a limited domain two radically different views may allow us to do pretty much the same. As a guide to navigation the astronomy of the ancients was not surpassed by the astronomy of Copernicus and Newton until long after Newton's death; and it remained competitive until the advent of recent technology. But according to the ancient view, the immobile Earth sat at the center of the universe while the stars were fixed on a gigantic sphere that rotated around the Earth. By keeping the stars in that sphere it was possible to calculate very precisely the position of the stars in the sky at any time of the year. And by reference to that position a sailor or an explorer could chart his course. In many respects it is still easier to apply the ancient view. In any event, to some extent the ancient and the modern views give us very similar practical guidance; they allow us, in a limited context, similar performances. But the views are not only different, they actually contradict each other: one forbids the motion of the Earth around the sun, the other requires it. If perchance we receive electromagnetic transmissions from another species, we should not conclude that those beings must have the equivalent of Maxwell's laws of electromagnetism. We may need Maxwell's laws in order to describe that those beings do. But their actual "laws," if they even think in such terms, may not be any more equivalent to Maxwell's than the Greeks' lack of motion of the Earth is equivalent to Copernicus' motion of the Earth around the sun.

I draw two morals from this discussion. The first is that there is no inevitable, no highly probable connection between the appearance of life and that of intelligence, nor between the appearance of intelligence and that of an advanced technological civilization. None of this rules out the possibility that extraterrestrial science exists; and even a small chance that it does may perhaps warrant a program to search for it, in spite of my remarks concerning the difficulties of communication. As important as these issues may be for the SETI program itself, I have not dealt with them in this paper. My more limited aim has been to use the assumptions behind the optimism prevalent in SETI to investigate the conditions that make human science possible. We have seen that those conditions are many, and that to a large extent they depend on the vagaries of natural history.

This brings me to a second moral, which I may only suggest here, although I have drawn it in some detail in the other chapters in this section.[6] The second moral is that science does not describe the world as is. Our own science does not allow us to discover reality in that sense any more than an alien science would allow an alien being to do the same. For it would be difficult to see why the point of view of any one intelligent species should be considered the way the world is when equally successful alternatives may have been developed in other worlds by other intelligent species. Why should our way of viewing the world be the right one? In our views of the world we find our stamp as much as we find the world's. Of course Kant had given us a similar insight. But now we can use natural and social history to explain why Kant was right.[7]

Notes

1. The conceptual underpinnings of this paper were first worked out in my book *Radical Knowledge: A Philosophical Inquiry into the Nature and Limits of Science*, Hackett, 1981 (Avebury in the UK). They are developed extensively in Chapter 8 of my forthcoming *The Dimming of Starlight*. Two other philosophers have taken up similar issues in a spirit much like my own. See Lewis White Beck, "Extraterrestrial Intelligent Life," Presidential Address, American Philosophical Association, December 1971, (Reprinted in the *APA Proceedings*, 1971, pp. 5-21); and Nicholas Rescher, AExtraterrestrial Science@, Chapter 11 of his *The Limits of Science*, University of California Press, 1984, pp. 174-205.
2. See for example Carl Sagan in *Communication with Extraterrestrial Intelligence*, Carl Sagan, ed., MIT Press, 1973.
3. Lynn Margulis, *Symbiosis in Cell Evolution*, W.H. Freeman Co., 1981. I do not wish to claim that symbiosis is the only way for complexity to arise. It is a plausible way, though.

4. Some of the points I make about natural history were also made by Stephen J. Gould in "SETI and the Wisdom of Casey Stengel," in his *The Flamingo's Smile*, Norton, 1985. They also appear in my *Radical Knowledge, op. cit.*
5. New findings suggest that proto-eyes are very ancient and thus that instead of convergence we have a case of common ancestry.
6. *Radical Knowledge, op. cit.*, Chapters 2 and 3.
7. I want to thank Jerry Cederblom for his many valuable comments.

3 Bohr and Evolutionary Relativism

There are two peculiar intuitions that afflict many philosophers. One is that realism must be right — even if they cannot prove it. Another is that relativism must be wrong — and they think they can easily show this. The purpose of this paper is to argue that these intuitions are misguided.

Behind the first intuition is the feeling that if realism is not right, pursuing science makes little sense. After all, the business of science is presumably to find out what is *out there*. If talking about what is *out there* is pointless (e.g., because realism is false or nonsense) then science has no particular significance. Popper, for example, talks about realism as a metaphysical presupposition of doing science.[1] Of course, trying to show that realism is true has been such a messy affair that many philosophers, particularly in this century, have achieved great sophistication in washing their hands off the issue. Nevertheless the hope of realism seems to come as standard equipment on most philosophers. Only realism, says Richard Boyd, can explain why scientific success is not a mystery.[2]

The surprising thing is that the most successful scientific field of the century is quantum physics, and quantum physics in its most orthodox interpretation is decidedly anti-realist. At least that is what Niels Bohr, the foremost thinker in the field, said explicitly: "...an independent reality in the ordinary physical sense can neither be ascribed to the phenomena nor to the agencies of observation."[3] By "phenomena" Bohr does not mean the sense data of philosophers but the *measured* subatomic objects. Phenomena are always the result of specific interactions with specific measuring equipment; but we should not thus conclude that they are two separate things, one of which gives us information about the other, for Bohr insists in the "*impossibility of any sharp separation between the behavior of atomic objects and the interaction with the measuring instruments which serve to define the conditions under which the phenomena appear.*"[4]

This interactionist view has many unpleasant philosophical consequences. One of them results from the fact that some measuring arrangements exclude others. In some an electron will behave as a wave, in others as a particle, but never as both. It all depends on what kind of experimental arrangement we employ, and thus we end up with *complementary* descriptions. *Real* things, however, cannot behave this way.

If we are realists we want to know *the* way the electron really is. These complementary descriptions, moreover, cover the gamut of Heisenberg's uncertainty relations. One experimental arrangement allows us to measure the momentum of a particle but it brings about an uncertainty in its position, and so on. Given all these considerations it seems unjustified to ascribe an independent reality to those subatomic objects. Furthermore, insisting on their independent reality requires doing away with the complementary of arrangements (and therefore of descriptions) that is inconsistent with that reality. But then we rule out finding out important aspects of the subatomic "realm." As Bohr puts it, "In fact, it is only the mutual exclusion of two experimental procedures, permitting the unambiguous definition of complementary physical quantities, which provides room for new physical laws, the coexistence of which might at first sight appear irreconcilable with the basic principles of science."[5]

Bohr thus made it clear that realist explanations in quantum physics preclude the ability to predict successfully in that area. That is, realism does not only fail to explain the success of quantum physics but the appeal to it hinders that success. I realize that many philosophers will claim that the matter is more complicated than I have made it out to be. Of course it is. Two paragraphs can hardly do it justice. And there are some who try to find some form of realism behind Bohr's words[6]; but I doubt that any amount of cleverness can twist those words to the point that they mean the opposite of what they so clearly say: "... radiation in free space as well as isolated material particles are abstractions, their properties on the quantum theory being definable and observable only through their interaction with other systems."[7] Indeed, for the reasons previously stated, such interactions entail "the necessity of a final renunciation of the classical idea of causality and a radical revision of our attitude towards the problem of physical reality."[8]

One important motivation for this attempt to give a "charitable" interpretation to Bohr's words is that they seem so obviously counter-intuitive to many philosophers, so at odds with the sophisticated things that most epistemologists think they have established. In the last few years the Aspect-type of experiments appear to have provided an experimental vindication for the Bohrian thesis that realism and quantum mechanics are mutually inconsistent. For some philosophers this result is a source of unhappiness, and although the topic in itself is very interesting, rather than add to the growing store-house of accounts of the EPR experiment, Bell's inequality, hidden variables, and the actual experiments themselves, I would like to explore the reasons for the now long reluctance to accept Bohr's words at face value.

Whereas charity leads those philosophers to suppose that Bohr did not mean what he said, great sympathy has instead greeted the realist position defended by Einstein, Bohr's main opponent in the dispute concerning the interpretation of quantum mechanics. This is not the place for an analysis of that dispute. But it is a good place to begin an examination of the metaphysical urgency so many feel in favor of realism. It is in part this: that the success of science is an extraordinary event that clamors for an explanation. Realism presumably gives science its due. Science works because either it gets at the truth about the world or at least approximates such truth. This of course makes science into a paragon of epistimic virtue, and therefore philosophy, epistemology in particular, would do well to learn the lessons that science has to offer. This may be seen as a proposal to "scientize" epistemology. Let us take it seriously.

If we want to find out what having knowledge of the world is, we should frame the issue within a scientific context, in this case a naturalistic context. *Someone* has that knowledge. That someone is a biological entity, subject to the workings of evolution. Its ability to know the world is thus a characteristic with which natural selection has endowed it. A characteristic, furthermore, that has evolved, as it is easily seen from the fact that different kinds of biological entities have varying degrees of ability to know the world. Taking the spirit of realism seriously, then, leads to a comparative study of the ability to know the world. The issue then turns on the manner in which natural selection endows diverse organisms with that ability. One suggestion highly sympathetic to realism is simply that those organisms who acquire ways to perceive the world as it really is (or whose perceptions approximate that reality) are in a much better position to survive. This suggestion presumably would not encompass theoretical knowledge in science, for it would be silly to claim that our brains were selected for atomic physics or even neurobiology. It might, however, encompass the human perceptual apparatus with which we make the observations that, if empiricism is right, will pass judgement on the worth of our scientific theories.

I do not wish my critique of realism to depend on this particular suggestion, which some may see as a straw man. More sophisticated versions of realism will be defeated all the same. For it is the most basic impulse behind realism that comes to grief once we begin to take evolutionary biology seriously. The most basic perceptual and conceptual equipment of organisms depends, at least in part, on their biology. An amoeba's perception of its environment depends on an interaction between its biology and that environment. A bird's perceptions depend on a central nervous system that results from two kinds of histories: the history of its ancestors as they coped with a long sequence of environments and its own individual history which fine-tuned its central nervous system. Perhaps some readers will be quick to point out that human beings have remarkable degrees of freedom in their central nervous system. I am inclined to agree, but a moment's thought will place our modes of thought back in a biological context. For our modes of thought also depend to a significant extent on our brains: People with different brain structures conceive of the world differently from most of us, as we can easily attest by ourselves during a visit to a neurological ward.

So far I have merely drawn some pretty straightforward consequences from the scientific realist's approach to epistemology, which begins by giving a nod of approval in the direction of an interactionist theory of knowledge (as it must if it is going to take science seriously, in this case biology, as it proclaims philosophy must do). In what follows I will proceed to draw some further, and very unpleasant consequences for the realist.

Now, different species may well (and sometimes do) experience the world differently. Those different ways of experiencing the world are relative to what we may call frames of reference (partially) determined by biology. It is as if the "world" were being measured in these different frames, and so it is not surprising that many different kinds of experiences are possible (as are ways of thinking about the "world").

It is obvious that some frames are better than others, at least for specific abilities. Some animals see much better than others, for example. And some, say, humans, are much more capable of producing powerful sciences than most if not all others. Perhaps it is tempting to think, as in the suggestion given above, that what makes one frame better than another is that it yields a more accurate representation of the truth, but biologically this makes little sense. A bird has a certain perceptual apparatus for the simple reason that it served its ancestors well enough. It succeeded, and because it did it was passed on. Another species may be served just as well by a different perceptual apparatus, which will also be passed on because it has enabled (or at least allowed) that species to succeed. These frames are the product of natural history, and natural history is a long path of millions upon millions of contingencies. One more accident here, one less there, and a particular lineage could have gone in a completely different direction, perhaps extinct. Moreover, no matter how good a frame is, natural history could have brought about another that is just as good, although very different. If, as a matter of fact, there is no other frame that enables another species to succeed just as well, neither here nor in any other part of the universe, then that is one more accident. Natural history could have made it otherwise.

An intelligent bat that had never seen birds or flying insects may think that his method of flight is *the* way to fly, nature's crowning glory. But wider experience, meeting hawks and honey bees, and reading the *Origin of Species*, soon convinces him that natural history not only could but has indeed produced alternative methods of flying which, it might be reasonable to add, are equally as "good."

The same intelligent bat might have also thought that his way of thinking about the world was *the* way of thinking about the world, until his new wisdom concerning different modes of flight cues him about the possibility of different modes of thought. The possibility, that is, that there might exist brains with structures as different from his as his flight mechanism is from those of birds and insects, but which nevertheless may serve those other species as well as his kind of brain serves his own species.

Our intelligent bat may then realize that his way of "viewing the world," his empirical science, even if it is the best that the collective efforts of batkind can produce, is no more successful than the possible "best" empirical sciences that those other species might develop. And then a terrible thought occurs to him: his science at its best was supposed to "represent" the world as it really was, that is presumably why it was so successful. But the truth of such representation of the world depended on its being a *unique* representation. He had imagined that any alternative view that could be considered true was not a true alternative but rather an *equivalent* representation (either logically, conceptually, or theoretically, in some cases also mathematically equivalent). He becomes aware, though, that views of the world may exist which are as good as his without being equivalent to it (not anymore than the birds' and insects modes of flying are equivalent to his). This shows him very clearly that even his best representation of the world (his science) need not be unique, no matter how good it is!

Matters look even worse to our bat when he ponders the contingencies of the *social* history of bat science — contingencies that greatly magnify the relativistic conclusions that already concern him, for the possibility is raised that even within the confines of bat science several directions could have been followed successfully. It is more proper, then, to think of the frames of reference as biologico-historical, although the epistemological lessons can be drawn from the simplified (i.e., mainly biological) account.

The most alarming consequence of this entire line of reasoning is that realism is wrong: science does not give us absolute truth, nor does it approximate absolute truth, nor can it aim to approximate absolute truth.

Before analyzing this consequence in greater detail, it pays to comment further on two aspects of the argument. What our bat's line of reasoning establishes is not that *all* (perceptual and conceptual) frames of reference are equally good, but rather that no matter how good a frame is, there *may* be some others that are equally good. Given that, it would be arbitrary to say of any one frame that it gives us *the way the world really is*. If it turns out that as a matter of fact there is only one frame that is good, we still cannot say that its view of the world is *the* correct way of viewing the world, for it is a mere accident that natural history did not bring about different but equally good frames. For the bat's view to capture the way the world really is, the world itself must cooperate by being one way and not many. But we have seen that many non-equivalent ways might represent "the world." The suspicion arises, then, that there is no truth of the matter to be represented here, since there can be no unique representation of the world. This last point is reinforced by the realization that the difficulty is not due to lack of information. For there is no new information that added to any one view could turn it into the unique representation of the world. And when every possible view will fail to represent "the world" correctly, we seem to face a dilemma: either the world is unrepresentable or else the expression "the world" is a mere convenience — there is no truth of the matter there.

Again, the first horn of the dilemma, that the world is unrepresentable, means not that the world cannot be represented at all, but that it cannot be represented "correctly" or "truthfully," at least as long as we understand such terms in their absolutist sense. Nevertheless many philosophers of science will hold on to their realistic instincts. They may argue, for example, that the case for different modes of thought is not like the case for different modes of flight. As suggestive as the analogy is, they might say, it does not rule out in principle that the very best frames will turn out to be equivalent, or, more subtly, that the job of the scientist-philosopher, as Clifford Hooker would put it, is to discover the invariants between those frames.[9] After all, their very success, and the fact that they are about equally successful (even if each does in its own way) cries out for an explanation. What other explanation can there be but that they all have captured some all-pervasive features of the world?

Our bat would find this reply most unsatisfactory on several levels. To begin with the presumed need for an explanation, he could point out that the worth of a frame is a function of its performance (or potential performance), but that two frames may be just as good in this respect without agreeing on the content. The beginnings of human science offer us a good example in this regard. The two-sphere universe of the Greeks was an extraordinary advance in navigational technology which sailors have preserved to this day. That is, in the area of navigation the performance of the Greek model was as good as that of Newtonian cosmologies. But the Greek model placed a stationary Earth at the center of the universe, two key beliefs denied by the Newtonian cosmologies. It is not obvious then that overlap in success needs to be explained by overlap in content (this point, incidentally, does in the notion that there must be convergence of content among advanced sciences — a notion that is already rendered suspicious by the flexible character of intelligence). As we will see below, it is success that explains truth instead. Our bat could next express puzzlement about the logic of the realist's reply: any theory of invariants would have to be developed within a particular frame, and thus natural (and social) history could have brought about a different theory of invariants ... and so on. How does the realist pretend to get out of his difficulty then?

More importantly, though, the realist seems to have missed the point altogether. If the bat's reasoning is correct, alternative frames based on significantly different brain structures are possible which may nevertheless allow the species in question to perform no less well. This is not merely a logical possibility: it follows directly from considering how natural history builds perceptual and conceptual frames. To deny this point is to assume that evolutionary biology is either incorrect or irrelevant — hardly the tack one would expect from a realist who is trying to scientize epistemology.

Moreover, here the bat can take a hint from Bohr. The skeletal and other arrangements that enable the bat to fly are incompatible with those that enable birds and insects to fly. Likewise, some cerebral structures enable certain modes of thought while ruling out others. Of course, humans whose cerebral structures vary significantly from those of the rest of the species may have an extreme difficulty trying to manage in the world. They are deviant. But in another species those structures, or even more unusual ones, may fit very well with the other characteristics of the species and thus be extremely adaptive and successful; just as skeletal structures or modes of respiration that would be great handicaps, if not outright lethal, in humans may well be extremely adaptive in birds or fishes. The views of "the world" produced by different frames may thus be *complementary* in a sense akin to Bohr's. It is possible then to produce information in one frame that is not logically, conceptually, theoretically, or mathematically equivalent to any produced in another, even if it is presumably about the same aspect of "reality" (they are "equivalent" only when that word is understood as a synonym of "analog," which is not the relevant sense here). As Bohr pointed out, one experimental arrangement that enables us to see the electron behave as a wave complements, but also rules out, another arrangement that enables us to see the electron behave as a particle. The wave and the particle descriptions are thus complementary — there is no sense in which they are *equivalent*. The same may obtain between descriptions produced in different frames.

Nor is there anything common to waves and particles for us to discover so as to provide a more complete description of the subatomic realm. As Bohr would argue, there is no information that we lack. Neither the uncertainty relations nor the notion of complementary are due to ignorance on our part. These are precisely the conditions described earlier in the chapter, conditions which led to the conclusion that since real objects could not behave that way, we should not speak of reality in the subatomic world. My suggestion is, not surprisingly, that since analogous conditions obtain in the case of the different conceptual frames, we should arrive at the same sort of conclusion: our talk of reality is misplaced — there is no truth of the matter "out there."

This is the view to which I arrived many years ago as a result of my failed attempt to develop an interactionist epistemology along the lines of Popper's scientific realism[10]. It was Paul Feyerabend who first drew my attention to the similarity between my evolutionary relativism and Bohr's epistemological position regarding quantum mechanics, a position that I had not fully appreciated until then. I do not mean to suggest, though, that Bohr's epistemology is on the whole similar to mine. The views of his that I favor were clearly confined by him to the description of the behavior of atomic objects, and he might have very well looked unsympathetically upon a generalization of those views to the entire field of empirical knowledge. He did try to extrapolate the concept of complementarity to a few other areas of experience, without much acceptance anywhere. The significant difference, it seems to me, is that the principle of complementarity made eminent sense where the classical notion of reality was found wanting by the quantum phenomena. In the philosophical view that I propose the classical notion of reality is found wanting even for the macroscopic phenomena. This finding requires a certain amount of reflection, and in that reflection an analog of Bohr's principle of complementary helps us understand the possibility of equally worthy and yet non-equivalent frames.

It appears to me that the main reason Bohr's view was found so counter-intuitive was precisely that scientific realism was deeply entrenched in our general epistemology and scientists (and philosophers) were loath to make an exception in the realm of the very small. As Einstein argued, no one is inclined to give up the program of scientific realism in the "macroscopic" realm, but, as he put it, "the 'macroscopic' and the 'microscopic' are so inter-related that it appears impracticable to give up this program in the 'microscopic' alone."[11] If, however, it turns out that for independent reasons we must give up the program of scientific realism in the "macroscopic" realm after all, and to give it up in favor of an epistemology congenial to Bohr's view about quantum mechanics, then it seems that Bohr's epistemology ends up looking very sensible indeed.

Many philosophers, however, would cringe at the thought that any form of relativism may be considered sensible. Their support for realism stems in part from the belief that the alternative to realism is relativism and that relativism is incoherent. The reason for the entrenched notion that relativism is incoherent is first found in Plato's argument to the effect that if we accept relativism then we must accept that all views are equally valid. And if all views are equally valid then we must accept absolutism as valid. But if absolutism is valid then we must conclude also that relativism is false. Thus if relativism is right it is false. Therefore relativism is incoherent. Nonetheless the first premise of this clever argument is based on a serious logical error. It is based, that is, on the belief that denial of the claim of absolute truth (relativism) implies that all views are equally valid. But that is not so. Such denial implies only that *there may be more than one valid view*. This is exactly what evolutionary relativism claims: that no matter how good a frame of reference is, there may be others that are equally good.

Another popular objection is that if we accept relativism there is therefore at least one truth to which we are committed: that relativism is true. Thus at the meta-level we revert to absolutism. Underpinning the objection is the belief that either the relativist is committed to the (absolute) truth of his position or else he is not serious in proposing it (and should not be taken seriously). But now we can see that this objection begs the question. The question is whether absolute truth should be our standard, and the absolutist refuses to accept any alternative which is not committed to absolute truth! In Chapter 1 I have shown that not only these but all the other (seemingly) reasonable objections to relativism found in the literature do not affect evolutionary relativism.

Nor is evolutionary relativism a purely negative philosophy; that is, evolutionary relativism need not limit itself to denying that absolutism (or realism) is correct. For within evolutionary relativism we find the resources to explain away the correspondence theory of truth while developing a more sensible notion of (relative) truth. In what follows I will attempt a brief summary (for more details see Chs. 1 and 4).

Let us return to our intelligent bat. He notices that the occasions in which he speaks of truth have some characteristics in common. When his sonar tells him that he is flying too close to a cave wall, he tilts his head to avoid a collision. His eyes, his taste buds, his jaw muscles all permit him to have a fully satisfying behavior toward the fruit he is eating. These are cases of perceptual interactions with the world that lead to behavior that can hardly be improved upon. These are, then, the best perceptions he can have. Even when he realizes that a different organism could have very different perceptions and still manage just as well, our bat would nevertheless have no reason to change his perceptions (even if he could) for they are the best perceptions *for him* in those circumstances. But in his pre-relativistic days the bat did not think about alternative perceptual frames of reference. On the contrary, the success of the behavior he based on those perceptions was such that he could not imagine how the world was other than the way he perceived it. He was thus more than strongly inclined to speak of truth on those occasions. Once he becomes a relativist, he still finds the ascription of truth useful: it helps him distinguish between those perceptions he does not want to change and those he would rather modify or replace altogether. They are still true, *but true relative to him* (or rather, presuming that he fits the norm, relative to his species' perceptual frame of reference). It is not truth that explains success, then, but rather success that makes it sensible to speak of truth.

Likewise at the level of science, a view may allow the bat such striking success in dealing with certain difficult questions that soon the bat adopts it as his means of viewing that entire area of experience (that is, it becomes his way of thinking about those aspects of the world, his paradigm — perhaps along Kuhn's lines). As success accumulates, the bat's view becomes so indispensable a pair of spectacles, that without it the bat can hardly make sense of the world. At that point it is extremely difficult for the bat to imagine the world otherwise. He is then inclined to speak of truth. Later, as a relativist, he will describe the matter differently. He will then realize that his interactions with the world might have been the best under the circumstances, but the circumstances change. They change because the very success of the view leads the bat scientists to employ it in new areas, where pronounced failure may now appear, or because the very refinement of the view changes the relationship between the bat scientists and the world, and so the criteria for what counts as success change. As in the case of perception, the notion of truth still provides some useful distinctions, but now he understands that the truth is relative to his biological frame of reference and to the historical circumstances. He also understands now a puzzling attitude of many bat scientists, namely the admission that, in the past, scientists had felt as confident about *their* views, and, what is worse, that future scientists may also feel just as confident about views that differ radically from the present ones. How could they speak of truth in the face of such an admission? It is understandable because what succeeded in the past is different from what succeeds now, as is what will succeed in the future.

To hold a view as (relatively) true is then to claim that it succeeds, that it is the best we can do within our conceptual frame, and that we cannot seriously imagine, let alone entertain, a different "picture" of the "world." This is the manner in which I hold as true the evolutionary biology I assume in my argument. It is also the manner in which I hold evolutionary relativism to be true. Bohr speaks of objectivity within an experimental arrangement, or within a frame of reference. The situation is analogous with this view of relative truth that I am sketching here.

I would like to close with another analogy. I have spoken of empirical knowledge as the result of interactions between the observers (or rather the measuring frames) and the objects of the world. In the macroscopic realm objects are composites, they exhibit many properties and many kinds of properties. A rock has a certain color and weight and resistance to pressure, and when you strike it with a shovel it makes an unnerving sound. Macro-objects are thus rich with potential interactions, and this richness permit us to think of those objects as continuous in time and space, for their properties are continuously coming to light, so to speak. The fabric of the macro-world thus looks to us as solid cloth. But as we examine this cloth with an ever more powerful microscope, gaps begin to appear in the cloth. At the micro level the kinds of potential interactions are greatly reduced. This poverty of interactions is the result of the very simplicity ascribed to the objects of the micro-world. We find ourselves in a situation in which certain types of interaction rule out others, and thus we come to a point in which, for some questions, we can choose only one kind of interaction. And thus there will be only one thing at a time that we can say about nature in those circumstances. To be able to say many things at once we need to turn the microscope around. But we should not forget that we are still looking through lenses.

Notes

1. K.R. Popper, *Objective Knowledge: An Evolutionary Approach*, Oxford University Press (1972), 203.
2. Richard Boyd, "On the Current Status of Scientific Realism," in *The Philosophy of Science*, eds. R. Boyd, P. Gasper, and J.D. Trout, MIT Press (1992).
3. Niels Bohr, "The Quantum Postulate and the Recent Development of Atomic Theory," in *The Philosophical Writings of Neils Bohr*, Vol.I, Ox Bow Press (1987), 54.
4. Niels Bohr, "Discussion with Einstein," in *Albert Einstein: Philosopher Scientist*, ed. Paul A. Schilpp, Open Court (1982), 210.
5. Niels Bohr, "Can Quantum-Mechanical Description be Complete?" reprinted in *Physical Reality*, ed. Stephen Toulmin, Harper and Row (1970), 139.
6. For example, Henry J. Folse, "Bohr on Bell," in *Philosophical Consequences of Quantum Theory*, eds. J. Cushing and E. McMullin, University of Notre Dame Press (1989).
7. Niels Bohr, *The Philosophical Writings of Neils Bohr*, Vol. I, Ox Bow Press (1987), 56-57.
8. Niels Bohr, "Can Quantum-Mechanical Description Be Complete?" *op. cit.* 132.
9. C.A. Hooker, "Between Formalism and Anarchism: A Reasonable Middle Way," in *Beyond Reason: Essays on the Philosophy of Paul Feyerabend*, ed. Gonzalo Munévar, Kluwer (1991).
10. G. Munévar, *Radical Knowledge*, Hackett (1981).
11. Albert Einstein, *Albert Einstein: Philosopher-Scientist*, *op. cit.*, 674.

4 Cultural Relativism and Universalism

Introduction

Most opposition to relativism is based on a very serious logical error: That in denying absolute and universal truth, the relativist is compelled to accept the notion that all points of view are equally valid. This error of logic has been accepted by the great majority of philosophers since Plato committed it. But it is an error nonetheless: the negation of the existence of an absolutely true point of view does not imply that *all* points of view are equally valid. It only implies that *several* points of view *could be* equally valid. It is incredible that for over a thousand years such an elementary point of the logic of quantifers has been ignored. Incredible or not, this historical situation has vitiated a great many philosophical disputes, especially those that have to do with science, society, or culture.

Let us see how. In many philosophical circles a distinction is drawn between science and culture. It is supposed that science searches for truth, for how things really are. But that reality cannot be the reality of the ordinary people: things just as we see them, hear them, and feel them. It is rather a platonic reality, one that hides behind the appearances. Thus the reality that science searches for is absolute and universal. This is not the case with culture. Different things, sometimes opposite things, are acceptable in different societies, in different cultures. And since opposite claims cannot all be true, it is clear that in matters of culture the acceptable does not coincide with the true. Universalism reigns in science, relativism in culture.

There are several ways of opposing this distinction between science and culture. The simplest one is to treat science as part of culture. This idea accords with Ortega y Gasset's notion that the culture of a society is comprised of its living ideas. Still better, we might say with Jesus Mosterín that culture is information transmitted by social learning. In both these senses it is clear that science *is* part of our culture. Unfortunately such a simple option only transfers the problem to another level: That there are some aspects of culture whose goal is absolute truth (science) and some others in which there can be no truth (social norms and art, for example). Consequently in this paper I will abandon Mosterín's sensible suggestion and will instead speak of science and culture as it is done on the streets of Philosophy: The first is an objective activity, the second is subjective.

Nevertheless it is possible to argue that such a distinction is illusory. For example, we might insist with Aristotle that art reveals to us a profound truth. Taking this position to an optimistic (and narrow minded) extreme, we may suppose that Western art, and only Western art, already captures those aspects of human experience which are universal (presuming also that Aristotle's profound truth is universal truth). This attitude puts in a position of inferiority artistic expressions whose roots are very different from those of Europe. Against it we find two strong lines of thought in the philosophy of this century.

The first is the analytic approach so prevalent in the Anglo-Saxon world, and specifically the notion that truth cannot serve as aesthetic criterion. For example, a poem may be masterful even though none of its sentences are true, whereas another one can be mediocre even if all its sentences are true.

The second line of thought stems from the celebration of cultural relativism: It is arbitrary and intolerant to think that outside of the Western world human beings have been incapable to express adequately their *own* experience. It is easy to demonstrate that many of our social and aesthetic beliefs, as well as many of our dispositions to judge in favor or against other beliefs, depend on the simple fact that we have been socialized in a certain culture and not in another. This fact creates the very strong suspicion that the (social and aesthetic) beliefs of other societies are as valid *for them* as Western beliefs are for Western society.

Unfortunately neither line of thought can avoid the disagreeable notion that the products of art, say, are not as worthy as those of science. This epistemological inferiority follows, in the analytic case, from the alleged fact that art has a purely expressive function (emotive), not a cognitive function. In the case of cultural relativism we have the obvious contrast with a science that searches for an absolute and universal truth.

The situation seems to be as follows: If there are universal truths or norms in culture, we have two options: They are to be found in one culture or in every culture (why not in *some* cultures?). If they are to be found in every culture it appears that we encounter great difficulties in discovering them. The variations from one culture to another are so many and so large that the universalist thesis seems to be completely in error. On the other hand, if we suppose that those universal truths already exist in the Judeo-Greek Western culture we have the no lesser difficulty of having to devalue the artistic expressions of the Orient or Africa entirely — to mention just one example — though they have been valued and even adopted by Western artists precisely because they were different from those known in Europe at the time (think of the influence of African ceremonial masks on Picasso's work).

It is possible to avoid this option, which seems as arbitrary as it is arrogant, and accept instead another: That (absolute) cultural truths and norms exist but we have no discovered them yet. Just as science has not discovered absolute truth yet, but progresses towards it, culture progresses towards its own absolute truth. Unfortunately this new option does not seem very promising. Besides, as we will see below, it militates against one of the most important objectives of culture, at least where art is concerned: to enrich our lives.

We find ourselves in a dilemma, then. On the one hand cultural universalism is not very plausible. But, on the other, if we accept relativism, which at first sight seems so reasonable, we apparently have to abandon the notion of cultural truth. We have to because of the paradoxes of relativism. If, as relativism presumably tells us, all points of view are equally valid, the universalist, or absolutist, point of view is valid too. But universalism implies that relativism is false. Therefore relativism is incoherent. This platonic argument reduces the relativist's "cultural truth" to a simple matter of taste.

But all this is an error, as we saw in the first paragraph of this paper. Relativism does not tell us that all points of view are equally valid. It only tell us that there might be several points of view that are equally valid. Consequently the cultural truth that relativism offers is not excluded, although of course it is a relative truth. This does not mean that it is not "real" truth (like that of science), for as we will see presently, the alleged "real" truth is also relative.

Relative Truth in Science

In its application to science, universalism is normally a form of realism. As such it advises us to take the scientific attitude very seriously. But the scientific attitude demands that we approach the question of knowledge with a naturalistic focus. That is, the scientific attitude demands that we consider how the knowledge of an organism emerges from its nervous system in interaction with the universe. This demand is very sensible. Organisms from very different species are capable of very different intellectual accomplishments. What they perceive, what they think, what they do with those perceptions can be very different also. It is no secret that the evolution of the human brain has placed within our reach a science and a culture (or cultures) that other animal species cannot even begin to understand. Even within a species biology can be very determinant. A person with cerebral injuries often perceives the world differently from the way other human beings do. Cerebral structures determine in great part an organism's modes of thought. Those modes of thought in turn present certain options and pose certain limitations to the possible development of science.

Now, our modes of perception and thought, as those of any other intelligent species, result from a long natural history that has changed direction millions and millions of times in accordance with a series of fortuitous accidents. It is supposed, for example, that a gigantic asteroid brought about the destruction and final extinction of the dinosaurs (and of seventy percent of all species) sixty five million years ago. This massive extinction created extraordinary opportunities for our mammalian ancestors. Without it we may still be little more than rats. Big accidents like this and small accidents that take place every day determine the future opportunities for evolution. The presence of a new fast predator changes the balance of the environment. It is not longer the biggest and stronger rabbits that survive but the fastest ones. Our brain structures — and thus our modes of thought — are the result of a long series of adaptations to a long series of environments. But just as we have arrived at a certain level of ease in our behavior with respect to the world, the forces of natural selection could have created other brain structures — that is, other modes of thought — of a comparable level even though they would be different from ours.

Let me explain. Birds' ability to fly is the result of a long evolution. But their mode of flying is not the only one nature has produced. Bats have a different mode of flight, and insects have an even more different mode of flight. It is possible, of course, that the universe does not contain brain structures comparable to ours, but that would be a pure contingency: natural selection could have produced others through different natural histories. Its failure to do so would be as much of an accident as its producing ours.

Different modes of thought lead to different ways of conceiving of the universe, that is, they tend to develop different sciences. There are those who think that the science of different intelligent species should converge, for although they are rooted in different modes of thought, they must ultimately take into account the same fundamental aspects of the universe. But this is an error (cf. Ch. 2). It is easy to see that the same aspect of the environment, for example a current of hot water, affects different species very differently: It kills some fishes, makes others more fertile, and is complete innocuous to still others. Convergence does occur from time to time, but it is in no way necessary. In the case of intelligent species two important reasons show that convergence is even more doubtful. The first reason is that one of the crucial characteristics of intelligence is plasticity, that is, the capacity to respond flexibly to the demands of the environment (for example, using imagination and memory to anticipate contingencies).

This plasticity of intelligence would ensure that species with different modes of thought may conceive of the universe (even if it is the *same* universe) in many ways — as long as it is advantageous to them. The second reason is that in the development of specific sciences, in addition to natural history, social history is also an important factor. Human contingencies, to put it that way, have also played an important role in the development of Western science. For example, only in certain social environments has it been possible to discuss and elaborate scientific points of view that caused doubt concerning the religious or metaphysical beliefs of the society.

What is the moral of all this? A scientific conception of the world does not happen in a historical vacuum, but is made possible by the interweaving of certain natural and social histories. If we concentrate on the natural aspect, we see immediately that our modes of thought are based on neural structures that result from a specific history in which randomness and opportunism have played a great role. No matter how effective that application to the world of those modes of thought, we should realize also that natural selection could have created other neural structures (or something to take their place) that would be no less effective. That is, whatever the "best" form of representing the world, nature could have created very different but equally effective forms. If so, we see that the "best" form of representation is not the only one that can be used to represent the world at that level of success. Thus it would be arbitrary to conclude that it is *the* true representation of the universe (or of reality).

If the "real truth" turns out to be relative to certain modes of thought historically given, we have two options. The first is to declare that truth does not exist. The second is to explore the sense in which it can still be called truth. The first option is not very sensible because there is an important distinction to be made between a point of view so convincing that we were ready to consider it as the representation of reality and other points of view that enjoy no comparable success (neither for us nor for any other species). Perhaps in the last instance we will decide not to call such a successful point of view "the truth," but it is worthwhile to explore this second option, although it should be clear that we are no longer using the word "true" in its absolutist sense.

Before carrying out this task, I should emphasize that, as I showed in Ch. 1, this evolutionary relativism does not fade before the traditional philosophical objections. Plato's main objection, that relativism is incoherent because it implies that all points of view are equally valid, is simply irrelevant. Evolutionary relativism does not imply such a thing. Another famous objection, more often encountered among physicists, is that if reality is relative to one point of view (or to the observer) we should then conclude that the universe was not born until there were beings that could observe it. This conclusion is absurd, of course, but be that as it may, it is not a conclusion one could draw from evolutionary relativism. What we want to say is that reality is relative to a historico-biological frame of reference, and that there might be several such frames that would permit as much success as the "best" frame that may occur to us. That is to say, such frames need not be concrete; it is sufficient for our purposes that they be potential frames. There is nothing strange in this explanation. For example, in the special theory of relativity, the mass and length of objects is relative to a frame of reference, as is time, but this is no reason to conclude that if there is no observer to measure such magnitudes, objects would have neither mass nor length and time would not exist.

We can also discard the favorite objection of philosophers: That if we believe in relativism then we would commit ourselves to there being at least one sentence that is absolutely true. That is, even relativists would have to be absolutists at the meta-level. But the only reason for thinking such a thing is that if we do not commit ourselves to the absolute truth of relativism, our proposal lacks seriousness (why do you claim *that* if you do not really think it is true?). When we evaluate this argument, however, we realize its fallacious, question-begging character: If someone casts doubt on absolute truth, the traditional philosopher demands that no alternative proposal be taken seriously unless it is offered as an absolute truth! (Or at least a candidate to such).

Although the evolutionary argument against absolutism could be limited to a *reductio ad absurdum*, we advance more the cause of understanding if we try to explain in which sense we accept the scientific considerations (especially those of evolutionary biology) that lead us ultimately to relativism. Of course, we could accept such considerations exclusively as the absolutist's premises, which we assume for the purposes of deriving a conclusion he will consider absurd. But this approach would force relativism to play a mere negative role.

When a scientific theory gives us a very successful way of interacting with the world we begin to see the world through that theory. The moment may come at which that theory becomes our scientific way of *thinking* about the world. It tells us what elements make up the world, what relations exist between those elements, and thus it sketches what problems we face and what types of solutions those problems may have (in the style of a Kuhnian paradigm). When this happens it seems to us very difficult, perhaps inconceivable, that the world may be otherwise. It is on such occasions that we speak of truth — the world *has* to be *so!*

A theory that has achieved this level of success has for scientists the same intuitive force that our most successful perceptions have in daily life. To leave the room I open the door, I do not attempt to go through the wall. I eat the apple but not the grass. Insofar as we can say that this theory, or these perceptions, gives us a "representation" of the world, it seems to us that such a representation is *unique*, that it is *the* representation of the world. This elimination of alternatives allows us to concentrate our minds and our behavior on a particular "gestalt," which often permits us a great ease in dealing with the environment, and thus it is only natural that we are inclined in its favor.

But when this natural inclination takes control of our philosophical investigations we put things backwards: We determine that the theory, or the perceptions, is true because is is correct (that is, *true*) and not, as we should say, that we believe it is correct because it is successful. An intelligent bat that had never seen a bird may find it natural to think that his way of flying is *the* way of flying. And perhaps he would also find it natural to conclude that his way of thinking is *the* way of thinking, and that the harvest of his thought about the world, his representation of the world, is *the* representation of the world.

Here we thus have a natural explanation of the force of our aspiration to demonstrate that our ideas correspond to the world: How can an idea represent the world correctly if not through a correspondence with it? Moreover, the "correctly" comes into play when we suppose that some representation is unique. This conjunction of philosophical notions, so innocent in appearance, seems to capture what we mean by "truth."

To arrive at this point, however, we forget many things. We forget, for example, that only in a metaphorical sense can we way that science represents the world. A theory gives us far more than a description of the world. It also gives us, among many other things, a way to interact with the world (at the very least: a way to facilitate posing questions to the world). Our description gains certitude when our resultant interaction with the world grows in sophistication and success. This point simply extends to science what we already know about our sensations in daily life. As Descartes pointed out a long time ago, our painful sensation of hunger need not resemble hunger in the least, what matters is that it makes us aware that we need to eat. A polished pebble can represent the sun perfectly in certain contexts, and the memory of an unrequited love may represent what my professional life means to me (although it does not resemble it at all). The value of a representation often depends on the practical context in which it operates — and science is not different in this respect.

As we can see, although correspondence with the world may seem to us even intuitive, a more ample understanding of science leads us to resist such correspondence. And an analysis of the historico-biological conditions of science makes us realize that we have put things backwards. As soon as we realize that science is a way of interacting with certain aspects of the world, the need for correspondence vanishes. It is possible, for example, even if such a thing as *the* world of the metaphysical realists were to exist, that a theory is successful because it allows us to brush up against the world in a fruitful manner, as a key brushes up against the keyhole in such a way that they complement each other and the door can now be opened.

Everything that can be said sensibly using the notion of truth as correspondence can be explained with this theory of relative truth. For example, we often accept a theory not because it has been successful but because we see great promise in it, or because our intuition tells us that the theory is going to give us the most acceptable way to think about the world — if we only take the time to develop it. This is the manner in which the majority of new theories begin to capture adherents, for otherwise all original ideas would stay on the drawing board forever. Only when the success of the theory is recognized will the almost complete confidence in the theory extend throughout the scientific community (there will always remain a few skeptics). It is only then that scientists will feel confident about the "truth" of the theory. The correlation between success and "truth" can thus be extended to all manner of epistemological contexts, and with it the confusion about what causes what.

But the theory of relative truth can also explain what remains paradoxical within the notion of truth as correspondence, namely, the scientists' acknowledgement that today's truth, like yesterday's, can be replaced in fifty years or less. The same certainty that is felt today was felt yesterday and will be felt tomorrow. Yet the theories are not the same. For modern philosophy this means that scientists are fools in epistemological matters because they easily coaxed. To make the situation even worse, they are cynical or irrational when they confess that, despite their certainty about the truth or their present theories, the science of the future could demonstrate that what is accepted now is little more than an illusion. But if we approach the matter from the biologico-historical point of view, we realize that no matter how good the interaction with the world that a theory may offer us at a particular moment, that interaction, which ultimately depends on human beings, can never be complete. Either the world changes of its own accord, as it has done, or changes as a result of our interaction with it, as it has also happened, or our theories change insofar as that interaction forces us to refine or replace them, which in turn changes our relation to the world. In any event, what at a given moment may seem impossible to improve upon substantially, in a new environment (new physically, scientifically, or both) may no longer serve as a model for interaction or behavior, and a new way of viewing the world becomes imperative. As long as success is the cause of our ascriptions of "correspondence," and thus of "truth," this notorious scientific attitude is easily explained.

In contrast with some pragmatic schemes, the dynamic process described here does not ensure a goal for knowledge (not even something like a limit towards which our science converges). A more fruitful analogy would be that of a horizon that recedes, or even better, several horizons that recede, for different natural or social histories may determine different modes of interaction.

What is the world really like then? Just as science tells us, of course — successful science in any case — just as the average scientist feels obliged to believe the world is like. But that representation of the world, however inevitable it may seem in its day, is *relative* to its time and circumstances, as well as to certain biological frames of reference. The "real truth," then, is not universal truth: It is relative truth.

This theory of relative truth can be discussed in greater depth, but that task is beyond the intent of this paper. I will mention, but only in passing, that other theories of truth do not offer much to the dispute surrounding our main topic. If we wished, we could employ this evolutionary theory of relative truth to explain easily, say, the theory of truth as coherence. We need only point out that a sentence about the world would be true if and only if it is consistent with a global theory that has the kind of success I have described. Besides, it would be very difficult to derive a realist thesis from a theory of coherence as truth (apart from the theory's internal problems).

Relative Truth in Culture

Given this theory of relative truth, we can now approach the question of cultural relativism from a wider and more fruitful perspective. The first point that becomes clear, of course, is that the presumed basic difference between science and culture is not justified: If absolutism (or universalism) is a requirement for speaking of truth, there is not truth in science (and probably none in culture either). But if we adopt a sensible position on this issue, we may then be allowed to speak of truth in both, albeit of relative truth. This position is quite congenial both to those intuitions that tell us that culture moves in the sphere of the relative and to those that claim that culture has important truths to offer.

It is advisable to qualify these remarks a little. The functions of art, to begin with an example, are far more varied than those of science. Art attempts to make us understand our emotional states or the relationships that exist in our social world, and it may also create in us new emotional states or new relationships. And it does all this sometimes in very indirect ways that exploit our emotional and intellectual capacities. But just as in science, different works of art may comply with their varied functions to several degrees of effectiveness (in a particular historical situation). That is, successful works of art capture our aesthetic imagination and give us a fruitful way of perceiving, thinking, and feeling about our relationship with the world, with our social environment, or with ourselves — or simply a new form of perceiving, thinking, or feeling.

The more effective the work (or type of work, as it happens, say, in the case of realism or impressionism in painting) the more difficult we think it is to ignore what that work reveals to us about ourselves or the world. This revelation can then come to seem to us to be the truth, a very profound truth in some cases. And here we arrive at a very important point in this essay: The truth that art reveals to us, the way of thinking that captures our imagination and we find so fitting, is often that of our own social or cultural environment. Or that chord that the work makes respond in our aesthetic being operates precisely because in the context of our social or cultural environment it makes sense. Clearly, then, artistic truth will be truth relative to a specific cultural environment, but it is not any less true for that, for the considerations that make us recognize it as true are of the same type as the relevant considerations in the case of science.

What we have just said about art can also be said about many of our other cultural activities. This does not mean that all works of art and all cultural activities reveal something to us, nor that all works of art are effective or significant. And it is certainly possible for us to make mistakes in our judgements about specific cases, just as we make them in science. What it does mean is that when art (or another cultural activity) is effective and significant *and* reveals something to us, then we feel the inclination to speak of truth — the same inclination we know from the case of science. And why not? It is one more case of relative truth.

Now, our cultural environments are not "complete" and cannot be completely closed; thus our artistic forms and other cultural activities are going to exhibit the same type of fluidity that science does, in many cases to an even greater degree than science. Art and culture are then going to change inasmuch as they search for their own horizons that recede, just as science does.

The pressure for change comes in part from a very important aspect of the nature not only of art in particular but of culture in general: that of enriching our emotional and intellectual lives. Art often takes us beyond our daily experience, or helps us discover an aspect of daily life that we had not noticed before. Art takes advantage, then, of new experiences or of new contexts for evaluating our experiences. Therefore the art and the cultural activities of other societies (past and present) can become sources of inspiration, of criticism, or even of affirmation.

We can show then that the insistence on universalism, or absolutism, in culture is an error no less serious, and probably greater, than the insistence on absolutism in science. Understanding human experience requires, at least in part, understanding the possible variety of human behavior in a variety of physical and social environments, many of which come to exist in all their complexity only because certain historical avenues of development were not closed, and still others owe their existence to inspired human imagination. Such functions of art, of culture, would be seriously curtailed if we deprive ourselves of the contributions of other perspectives.

Of course it is worthwhile to search for what we have in common, as it is to discover our differences. Both are part of our human situation, and therefore they deserve our curiosity. But what we have in common can be expressed in many different ways, and even if it were expressed in only one way we should not speak of absolute truth. Just as in science, at most we would have obtained a relative truth.

In coming to the end of this essay, I should clarify once more that evolutionary relativism does not imply that all cultural forms are equally good. Some will be more effective than others in carrying out their functions. Some art opens new horizons for us, other art only repeats trivialities in a trite manner. But there is always the possibility that alternative cultural forms might be as effective as the best that we may conjure up at any given time, that their contributions might be no less valuable, no less worthy of being called "truth." Relative truth.

References

Jesus Mosterín, *La Filosofía de la Cultura*, Alianza Editorial, Madrid, 1993.
Gonzalo Munévar, *Radical Knowledge*, Hackett, Indianapolis, 1981.

5 A Note on Margolis

I find myself very much in agreement with many of Joseph Margolis' pronouncements in defense of relativism. In his *Pragmatism Without Foundations*,[1] he is quite right in insisting that relativism need not be construed as a thesis based on the 'cognitive facilities' of man. As he says, we may 'hold that men do not or cannot truly know the actually structure of things ... because on our best theories, there is no such unique and independent structure to be known'. (p. 54) I am also very sympathetic to his attempt to demonstrate that relativism offers the philosophical advantages that have generally been ascribed to realism; and I was particularly pleased to see Margolis' insistence in making sense of science on relativistic terms.

In general, I find throughout Margolis' early chapters many echoes of themes that I have developed at great length in my own work. And so it is not surprising that I would be quick to praise him. Nevertheless, I had great difficulty in determining Margolis' general position, in spite of my admiration for several passages. It seems to me that in the long run Margolis sells relativism short.

My first difficulty comes in his distinction between internal and external relativism. Internal relativism is supposed to be sensible relativism, the kind that thoughtful realists could live with. External relativism has an irrational streak and is therefore bad advertisement for a position — relativism — that deserves much better. Margolis' favorite example of external relativism is provided by the common themes of Feyerabend and of Kuhn (in his naughty days). According to Margolis, the mark of external relativism is that it permits no method by which to decide between competing paradigms or theories. Internal relativism, on the other hand, allows us a (weak) methodological basis 'for epistemologically supporting or confirming claims [about the world]'. In some cases the decision may have to wait for the future development of science; but the mere fact that a question cannot be decided now, does not show that it can never be decided.

Now, what Kuhn and Feyerabend have argued is that there is no method that can be applied in all circumstances to yield progress. More strongly, Feyerabend's position is that all methods have exceptions and that on occasions the good of science requires that method be violated. But both Kuhn and Feyerabend allow that under certain circumstances some decisions will be more appropriate than others. We may say that one idea, in a particular version, fits the exigencies of research better than a second idea, in a particular version. But we cannot exclude the possibility that a different version of the second idea may later prove superior. In the case of the Copernican revolution, for example, method would have apparently required that we do not entertain seriously, let alone accept, a view that conflicted with the facts. But the view that Earth did not move and was the center of the universe, which Aristotle and the ancient astronomers had good reasons to reject, made a comeback even though it was in conflict with 'plain' facts. (See Part II and Appendix B). It did so because those facts needed to be interpreted in the light of certain theoretical assumptions. When those assumptions were challenged, so were the contrary facts (e.g., that a stone dropped from a tower falls straight down). It is precisely the breakdown of the separation between theory and fact that permits the challenge of any decision based on any empirical method (since presumably what makes it empirical is that it stresses the primacy of facts over theory).

A challenge is thus always permitted, as far as epistemology is concerned (for Feyerabend, that is: the naughty Kuhn used to require a crisis). This is not to say that at any time and under all circumstances all views are equally good whether any one has ever taken the trouble to develop them or not. It is not, then, that no views can ever confront one another in 'an evidentially pertinent way' (p. 56), but that what counts as evidence may come under theoretical challenge and may thus also become part of the confrontation.

In contrast to this 'irrational' position Margolis offers internal relativism. He illustrates what he has in mind by 'weak methodological means' by pointing to Quine's indeterminacy of translation. Margolis is particularly struck by Quine's application of Duhem's thesis to translation, that is by the claim that a speaker's dispositions to verbal behavior can be accounted for by many alternative sets of sentences. By altering those sets of sentences 'the divergences can systematically so offset one another that the overall pattern of associations of sentences with one another and with nonverbal stimulation is preserved'. (p.57). This presumably is internal relativism. The striking difference with external relativism is that even though many alternatives are permitted, some others are ruled out because they no longer cohere with nonverbal stimulation. Going back to science through Duhem, I presume that the internal relativist would permit several alternative to co-exist, but not those that conflict with the plain facts (the counterparts of nonverbal stimulation, I take it). But then the difference between external and internal relativism is simply that external relativism is better informed about the history and practice of science. Once we get away from caricatures, Kuhn and Feyerabend are as 'internal' as Quine, except that they have given strong arguments against the demand for the agreement with the facts that Margolis praises in Quine (what Quine himself would say is another matter). Perhaps Margolis has strong counter arguments. In that case it would be nice of him to offer them. Or perhaps I misread him altogether: something else is what always permits us to eliminate some alternatives while keeping several others. If Margolis knows what that something else is, by all means he should confide in the rest of us. Rationalists and objectivists of all sorts have been desperately looking for it since the early sixties.

    I do not expect that the so-called external relativists would be much impressed by Margolis' point that questions that seem undecidable today may be decided by future science. For all this point amounts to is that within a future theory or paradigm certain questions will be decidable. But neither Margolis nor the extenral relativists, expect to arrive at that future theory or paradigm by some sort of decision procedure — the problem of method still remains. Even hindsight has its troubles, for its perspective is as much the result of relativistic processes as the controversies upon which it bestows its enlightenment. An exhaustive discussion of other illustrations of his distinction would not serve the purpose of this short piece, since I find them even less promising. Among those illustrations I include his erroneous treatment of the issue of incommensurability. (For this issue, see Appendix B).

    I would like to move on instead to the second major difficulty I see with Margolis' internal relativism. His compromise with realism suggests that his view is simply one more form of realism — one that overlaps with Putnam's internal realism and which in many respects is not as sophisticated as Hooker's causal realism. Hooker recognizes that it would be silly to imagine that at this stage of human inquiry we should have gotten the right story of what the world is like.[2] Not only are the specific claims of science up for challenge, so ar the very procedures by which we try to gain knowledge about the world. Evidence, method, and epistemology are at a primitive stage. To advance in all those fronts we need different points of view — pluralism. And at any stage of inquiry even the most comprehensive point of view will be only part of the story, influenced by the sorts of historical and social considerations that Margolis brings up. What presumably makes Hooker a realist is that those scientific points of view are the result of interaction with the world, and however clumsy, they are thus models of what the world is 'really like'. Even wildly different models may still be models of the same thing (as the same emotion, say love, may give rise to entirely different expressions of it in a painter and a music composer, in this case the same world gives rise to different models of it). Since it may well happen that no model will ever be completely satisfactory, and thus we may never be in the position to exclude alternatives to it, we may never know what the world is really like. But this is not to say that there is not truth of the matter, it is only to say that at many stages of inquiry we might not be able to say much about the truth of the matter. Since I find it very difficult to tell how Margolis' view differs in kind from sophisticated versions of realism such as this, I find it difficult to call him a relativist. Perhaps his view, as Hooker's, is a relativized realism.

My previous remark about Margolis' failure to draw a distinction between internal and external relativisms should not be taken to indicate that external relativists, such as Kuhn and Feyerabend, are somehow committed to realism. Margolis' internal relativism is a mere instance of philosophical wishful thinking. If his wishes could come through, however, then his position would be one more form of realism. In relativism proper, by contrast, the line is drawn clearly: ' ... on a ... contested point, there simply would be no truth value or truth-like values *to be assigned at all* ...' (p. 58) The 'at all' phrase holds for relativists as long as truth values are understood in an absolute sense. All relativists would permit the relativization of truth – if anything, that is one of the main reasons they are so infamous. Thus the 'internal' assignment of truth values (internal to a theory or paradigm) is not going to bother the 'external' relativist. The relativists insist that absolute truth values cannot be assigned precisely because they agree with Margolis that there is no 'unique and independent structure [of the world] to be known', even on our best theories (and the relativists would add, even on our best future theories, or in the best theories of gods and titans). Any relativist that holds back on this point is simply a misguided realist. He thinks he opposes realism, but he opposes only the vapid versions of realism that abound in today's literature. If this so-called relativist were to examine sophisticated realism, he might find little quarrel with it.

As far as I can determine, Margolis is not of one mind on this subject. Perhaps his willingness to compromise led him to disown, at least temporarily, his own description of what makes a relativist. He could have stuck to his relativistic guns by realizing the full consequences of his own insight about the priority of praxis with respect o human speculation: ' ... that theory, science, cognition itself are guided by the largely tacit, biologically grounded activities of human societies seeking to survive and reproduce their numbers, always in accord with the contingent pattern of life of particular cultures'. In any event, this is the route that I have favored in my own work.

Although Hooker's realism goes a long way to assimilate the advantages of my own interactionist epistemology of science, and though he also favors a biological approach not unlike my own, on the crucial point between realism and relativism we still disagree. To the extent that coming to have a certain point of view depends on biological and social factors, it also depends on the natural and social history of that individual, and that individual history ought to be understood in the appropriate biological and social contexts, that is in terms of the social and natural history applicable to that individual as a member of a group and a species. As soon as we realize this we realize that natural and social accident have much to do with the formation of a particular point of view. From our understanding of natural history, for example, we know well that nature does not seek out goals or try to fulfill archetypes. Natural history could have gone in a great variety of ways even for intelligent beings like us, and it is sure to have gone differently for other intelligent beings anywhere else in the universe, if any such do exist. There is, furthermore, no preference for one way over another, except for a long sequence of successes in a variety of environments. But for any such sequence natural history could have provided another (even if in fact it has not). Beings with high cognitive abilities are also the result of such paths of natural history. And thus even if we consider theories developed when our cognitive abilities are working *ideally*, we may pause to consider that natural history could have developed a very different path to cognition which may be as good as ours (and if we are not the best example, let us take titans at their ideal best: the same points about natural history will apply to them — they do to God as well, which actually goes to show that no being can be omniscient in the standard sense of the word; but that is the subject for another paper).

Now, if there may be several equally good alternative ways of understanding nature, it seems pig-headed to argue that any one of them is the right way. This indicates to me that no one description is the description of the actual structure of things, no matter how successful, since that success is always in comparison to other actual alternatives and it is a matter of accident that the one to come out on top did not actually have stronger competition. But if no description can be the description of the actual structure of things, I wonder in what sense we can speak of such a structure. Realism ends by making reality ineffable. I do not think that it helps much to say that the alternative understandings of nature produced by the top alternative forms of cognition are all 'models' of the actual structure of things. Models of what? If we could say of what we would have the ideal description. But that is just what we cannot say. For the relativists the matter is not at all that complicated. Once truth and reality are relativized, we can always tell you what reality is like. At this stage of inquiry we may have to take back later everything we now tell you about the structure of things. But under the same ideal circumstances in which realism failed, we would not have to take back anything.

This evolutionary relativism can be easily defended against all the common objections, but the interested reader will have to look at the longer expositions of it in Chapters 1, 3 and 4.[3] As I close this note, I must confess to some uneasiness about the extent of my disagreement with Margolis. The presentation of his argument was so complex that I may not have been up to the task of understanding it fully.

Notes

1. Margolis, J., *Pragmatism Without Foundations*. Blackwell, Oxford, 1986.
2. Hooker, C.'s realism is developed in his, *A Realistic Theory of Science*, SUNY Press, 1988.
3. See also Munévar, G. *Radical Knowledge: A Philosophical Inquiry into the Nature and Limits of Science*, Hackett, (Avebury in the UK), 1981.

PART II
EVOLUTION AND RATIONALITY

6 The Connection Between Evolution and the Nature of Scientific Knowledge

If the mind can be said to be a product of evolution, and since science is a product of the mind, it is worth considering what connection, if any, exists between scientific knowledge and evolution. This point refers not to the knowledge of evolution per se, but rather to whether scientific knowledge is somehow a result of evolutionary pressures (that it has adaptive value, say) or that evolution presents the key in understanding the nature of scientific knowledge. Proposals along these last two lines have been made by thinkers like Ernst Mach, Konrad Lorenz, Karl Popper, and others. It is the purpose of this chapter to review the most important points of such proposals and to present a case for the connection between scientific knowledge and evolution. I will begin this paper by considering two good ways of not looking for such a connection.

In the 19th century, thinkers like Mach, Spencer, and Poincare began what has been termed "evolutionary epistemology." In his contribution to this field Mach made two important claims. The first is that science has the function of reproducing facts in thought in order to save, or replace, experiences.[1] The second is that through evolution the mind adapts itself to the world.

> It is not to be denied that many forms of thought were not originally acquired by the individual, but were antecedently formed, or rather prepared for, in the development of the species ...[2]

This line of thought was captured in a nutshell by Spencer, who said, "What is a priori for the individual is a posteriori for the species."

If Mach and Spencer are correct, furthermore, by adapting to the world the mind comes to reflect it (in loose terms "the structure of the world forces itself upon the structure of the mind").

Not every evolutionary epistemologist agreed with the realism implicit in this view, however. Poincare, for example, felt that we choose our theories not because they are true but because they are more convenient.

> It has often been said that if individual experience could not create geometry the same is not true of ancestral experience. But what does it mean? It is meant that we could not experimentally demonstrate Euclid's postulate, but that our ancestors have been able to do it? Not in the least. It is meant that by natural selection our mind has adapted itself to the conditions of the external world, that it has adopted the geometry most advantageous to the species; or in other words the most convenient. This is entirely in conformity with our conclusions; geometry is not true, it is advantageous.[3]

There are many rewards in this 19th century approach. Poincare, again, provides a nice synthesis to the rationalist-empiricist controversy:

> We see that if geometry is not an experimental science, it is science born apropos of experience; that we have created the space it studies, but adapting it to the world herein we live. We have selected that most convenient space, but experience has guided our choice; as this choice has been unconscious we think it has been imposed on us; some say experience imposes it; others that we are born with our space ready made; we see from these preceding considerations what in these two opinions is the part of truth, what of error.[4]

As interesting as this 19th century approach may be, it is marred by its tendency to tie scientific knowledge and evolutionary pressures much too closely. For instance Mach claims that:

> We are prepared, thus, to regard ourselves *and every one of our ideas* as a product and a subject of universal evolution; and in this way we shall advance sturdily and unimpeded along the paths which the future will throw open to us.[5] (My emphasis)

Of course, the problem here is that our scientific ideas have changed quickly and considerably for the past few hundred years. Thus, if they were biologically embedded, as Mach suggests, our biology must have changed in a similar fashion. But that is clearly not so. Perhaps he would have been wiser, or at least safer, to restrict the evolutionary claims to the "higher forms of thought" of which the early evolutionary epistemologists spoke so often — and then to suppose that those higher forms of thought, or mental "categories," constitute the limits within which our scientific theories can operate. But then specific scientific theories would not be the result of natural selection, no matter how general their character, even if the higher forms of thought are.

It is certainly possible to restrict oneself to tracing the evolution of the structure of the mind (this would be evolutionary psychology), and leaving the problem of science for later. Indeed, such a task may be of great value to epistemology. This has in fact been the approach taken by Konrad Lorenz, one of the greatest pioneers in the field, although he has also come very close to dipping his toes in Machian waters.[6]

A second approach to evolutionary or biological epistemology severs the connection between scientific theories and survival value altogether. Such has been the tack taken by Karl Popper and Stephen Toulmin.

In places Popper seems to have much in common with Mach. He talks, for example, of exosomatic evolution (theories are like organs that we develop outside our skins).[7] The greatest advantage of exosomatic evolution is that we let our theories die in our stead. But in other places, Popper denounces the resemblance and proposes instead "something like a refutation of the now so fashionable view that human knowledge can only be understood as an instrument in our struggle for survival."[8] He disclaims any such interpretation of his position.

> ... I did not state that the fittest hypothesis is always the one which helps our own survival. I said, rather, that the fittest hypothesis is the one which best solves the problem it was designed to solve, and which resists criticism better than competing hypotheses.[9]

Popper's suggestion is rather that the theoretical adaptation he has in mind is adaptation to an "objective realm" separate from the world of things (i.e., from the universe), that is, adaptation to what he calls the "Third World."[10] A rather similar approach, but without the entanglements created by the proliferation of separate worlds and the like, is provided by Stephen Toulmin. Toulmin's position is actually worked out in far greater detail.[11] He, as Popper, argues that a theory should be preferred if it adapts best to the intellectual environment it faces. This environment is provided by the ideas, techniques, and problems that the scientific community of the time finds pressing and important. Toulmin requires mechanisms for variation and selective perpetuation (generation of alternative views, and what Feyerabend calls the "principle of tenacity," although applied only to successful candidates). For these mechanisms to operate, there must be a forum or a court in which the new alternatives may be "heard," and a tribunal that will preserve the accepted view until one of the alternatives can show that it is better adapted (or perhaps, adaptable) to the discipline's intellectual "environment." This tribunal is rather formally constituted and it requires a professional society for its existence. The existence of such tribunals assumes great importance for Toulmin, and becomes the main element that differentiates scientific from other disciplines.

There are two problems with this approach, however. The first is that accounts such as Toulmin's or Popper's can be called "biological" or "evolutionary" only in an analogical sense. And I am just not sure of the justification for the analogy. This may be a minor problem, though. The major problem is that this approach fails to draw as sharp a distinction as required between scientific and non-scientific intellectual activities. It seems, for example, that if ethics and art cannot already be considered scientific in Toulmin's account (I think, for instance, of the French Academy of last century), they could in principle become so. Is it reasonable, however, to think of ethics and art as sciences? It seems more reasonable to suppose that if they fit Toulmin's account they should at most be considered rational. (Even this suggestion is besieged with difficulties, for it is less than clear that fulfilling Toulmin's account suffices for rationality — but I will not go into this additional issue here.) But we should not expect that every rational enterprise be a scientific one as well. More is needed. What is this more?

Before answering this question it may be worthwhile to step back and consider in more detail why one should think that there is a connection between scientific knowledge and evolution. To begin with, I do not think it is that controversial any more to say that intelligence is a product of evolution. If intelligence is such a product, it is related to other biologically based structures, then, although there are some fundamental differences. In Piaget's words, intelligence is "the form of equilibrium towards which all the structures arising out of perception, habit and elementary sensori-motor mechanisms tend."[12] Intelligence is not of "a kind" with other structures, then, for it is distinguished by its much wider scope of application in time and space.

> [Intelligence] is the most highly developed form of mental adaptation, that is to say, the indispensable instrument for interaction between the subject and the universe when the scope of this interaction goes beyond immediate and momentary contacts to achieve far-reaching and stable relations.[13]

This operation of intelligence on the environment, even though fundamentally removed from it in space and time, agrees pretty much with the findings of neuroscience. Indeed we often find the growth of long and complex neural circuitry associated with the development of intelligence.[14] I place emphasis on this entire matter because it will become quite important shortly. It would be wrong, however, to conclude that science has adaptive value just because intelligence does. The functions of biological structures are not limited to those for which they were selected. Thus more of a motivation is required to find a link between science and evolution.

Now, it seems to me that the intelligent mind interacts with the universe by forming views of it and then trying them out. This accords very well with the contemporary philosophy of science developed by such thinkers as Kuhn, Feyerabend and Lakatos, all of whom would claim in some form or another that our scientific views structure the very manner in which we experience nature.[15] It also accords with R.L. Gregory's remark that "science is the cooperative perception of the universe."[16] If this sort of position is correct, and I think that it is – although this is not the place to argue in its behalf – then we can justify the connection between science and adaptive value via intelligence. If intelligence has adaptive value, in part precisely because of the way it interacts with the world, and if science constitutes the means to that interaction, then science has adaptive value as well.[17]

What should be expected of science if it truly has an evolutionary character? A science with such character would permit us to get along in the universe, generally, in changing combination of the following three ways:

(1) dealing with greater ease with our environment (our "niche");
(2) increasing the number and diversity of environments that we can deal with (enlarging the "niche");
(3) coping with a continuously changing environment (which puts a premium on flexibility or response).

It is by no means far-fetched to suggest that science does all these things for us. For example, it places at our disposal the technological means to lead more comfortable lives; it allows us to take advantage of the resources of ocean floor, air, and even of worlds beyond our home planet; it provides a larger picture (in time and space) of our habitats, and alerts us to the direction of change. None of this goes to show that science guarantees better adaptation, only that it makes adaptation possible if we choose to employ it wisely. But then adaptability should be understood in terms of potential, not actual adaptation.[18]

I would like to leave at this point the positive side of the case, i.e. of the reasons that there might be to expect a connection between science and evolution, and concentrate instead on the negative side, i.e. on what many think are the very powerful reasons against such a connection. The strongest objection is that a scientist is seldom concerned with the survival value of his theories when doing science. What motivates a scientist is the desire to get to the truth, his intellectual curiosity. The same point applies to the acceptance of theories: The scientific community prefers (or should prefer) those theories which best satisfy its intellectual curiosity, i.e., which are nearer the truth, independently of their practical applications. In fact, it is often impossible to tell what practical applications, if any, a particular hypothesis, theory, or even branch of science will have. But without such practical applications there could be no connection with survival value.

Of course a scientist may be motivated by truth or intellectual curiosity. But he may also be motivated by money, fame, the desire to impress his former teachers, and by many other things as well. In any event, the objection exemplifies too narrow a view. The assumption made by the objection is that survival value is more or less connected with foreseeable application. It is often said, for example, that whereas animals can only take care of immediate and pressing problems, i.e., react to them, we can behave in ways that do not constitute a reaction to any compelling demands of the environment, we are endowed with curiosity (a higher form of which provides much of our scientific motivation), and curiosity liberates us from the drudgery of "plain" animalhood.[19]

In reply to this objection we can make two points. The first, which has been prepared beforehand, is to show how unfounded the crucial assumption is. Such is the assumption that survival or adaptive value involves immediate or foreseeable applications. Let us remember that even though intelligence need not enjoy such applications it is recognized to have adaptive value. Just as the objection fails against intelligence, it should fail against science, one of the tools of intelligence. Indeed if my description of what science does for us is correct, we could speak of adaptive value in the long run.

Still there is a long standing prejudice that what distinguishes science form other activities is that science tries to force upon itself the verdict of experience (through predictions, testing, and so on). Of course science does connect with the world. But the connection is often much more subtle and indirect than the long standing prejudice indicates. In a different context I have developed the following three criteria; together they indicate the nature of the proper connection between science and experience.[20]

> Alpha': The theory enables a society, presumably as represented by its scientists to deal (directly) with the universe (or rather with a portion of experience). This involves predicting, retrodicting, etc.

> Beta': The theory structures experience, or develops the theoretical or experimental tools required for the structuring, such that the experience can be better placed in the "causal" network provided by the general theory. This comprises the necessary mathematical, theoretical, or experimental work to fit the general theory to the world, and amount to what Kuhn has called "the articulation of a paradigm."

> Gamma': The theory connects experience (the environmental input) to the rest of scientific knowledge (bringing it into the causal network perhaps for the first time). The fulfillment of this criterion indirectly allows the society to deal with the environment. Darwin's theory of evolution might be an appropriate example.

It can be seen then that the beta' and gamma' criteria, which provide for the refinement or incorporation of theories of more direct application (potentially), constitute indirect but secure links between a species' scientific understanding and its universe. More specifically, it can be seen how they contribute to dealing with greater ease with our environment (our "niche"), increasing the number and diversity of environments that we can deal with (enlarging the "niche"), and coping with a continuously changing environment (this contribution is, again, indirect — as is that of intelligence in general).

The second point of the reply undermines the presumed conflict between the role of intellectual curiosity and the possible adaptive value of science. Konrad Lorenz has shown that curiosity not only exists in animals as "low" as the Norway rat and the raven, but it is also of great survival value. The satisfaction of curiosity allows such species to "construct" their environments for themselves, and thus to exercise a great versatility of application due to their minor degree of specialization (they are equipped with very few and very broad releasing mechanisms and not many innate motor patterns). As a result, such species, which Lorenz calls "specialists in non-specialization," can adapt to a great variety of environments. This is partly the key to the survival value of curiosity. Homo sapiens would be, in Lorenz's account, specialists in non-specialization par excellence.[21] And curiosity, of which intellectual curiosity is a form, would serve as a prime motivator in getting the species to exercise its intelligence.

I must caution, nevertheless, against a wrong way of handling this matter of the adaptive value of curiosity. One might be tempted to claim, as Popper does, that science has its origin in problem solving. Scientific curiosity would thus be directed by responses to difficulties generated by the environment (initially, at least).[22] This account would leave no room for what may at first appear to be "useless" research (and thus it would fail to include much theoretical research). Once again it pays to draw a parallel to the animal kingdom. In those animals that carry out active investigations of the environment, that is, that try to satisfy their curiosity, it is quite evident that such investigations seldom are direct responses to environmental needs. According to Lorenz:

> The young raven conducting its 'investigations' is not motivated to eat, and in the same way a young Norway rat repeatedly dashing back to the entrance of its retreat from various points within its range is not motivated to flee. This very independence of the exploratory learning process from momentary requirements, in other words from the motive of the appetite, is extremely important. Bally (1945) regards it as the major characteristic of play that behavior patterns really belonging in the area of appetitive behavior are performed 'in a field released from tension.' As we have seen, the field released from tension — a sine qua non for all curiosity behavior just as for play — is an extremely important common feature of the two kinds of behavior![23]

Of course we can expect great differences between the curiosity behavior of animals such as ravens and that of man. The difference lies in the fact that man's investigative behavior is pursued until the onset of senility, a fortunate characteristic made possible by the neotenous nature of our species. In other animals such investigations are restricted to an early phase in individual development. Curiosity ends when play behavior ends.

If science is an attempt to satisfy intellectual curiosity, it seems that its origin is not to be found in problem solving but in play! Its preservation, furthermore, seems dependent on the very happy accident that we are able to keep our childlike sense of wonder.

Conclusion

An examination of the nature of scientific knowledge shows that there are good reasons to expect a connection between science and evolution. The main objections against this position fail, thus strengthening the case for the evolutionary connection.

Notes

1. According to Mach, one of the main functions of science is the economy of thought. Mach's biological or evolutionary theory of knowledge can be found in his *Popular Scientific Lectures*, Open Court (1943), pp. 186-235, and in section IV of chapter IV of his *Science of Mechanics*, Open Court (1942), and p. 222.
2. *Popular Scientific Lectures*, op. cit., p. 222.
3. *The Foundation of Science*, The Science Press (1946), p.91.
4. *Ibid.*, p. 428
5. *Popular Scientific Lectures*, op. cit., p.235.
6. See especially his *Studies in Animal Behavior*, Harvard University Press (1971).
7. See, for example, his "Is There an Epistemological Problem of Perception," in I. Lakatos, A. Musgrave, (eds.), *Problems in the Philosophy of Science*, North-Holland Publishing Company (1968), p. 163.
8. *Objective Knowledge*, Oxford University Press (1972), p. 264.
9. *Ibid.*
10. *Ibid.*, p. 106.
11. Especially in his *Human Understanding*, Vol. 1, Princeton University Press (1972).
12. *Psychology of Intelligence*, Littlefield, Adams & Co. (1966), p.6.
13. *Ibid.* p. 7.
14. A key is the insertion of delay and longer-circuiting pathways, in coordination with memory. See, for example, Robert B. Livingston, *Sensory Processing, Perception, and Behavior*, Raven Press, (1978), p. 6-18.
15. See for example, T.S. Kuhn, *The Structure of Scientific Revolutions*, 2nd Edition, University of Chicago Press (1970), P. Feyerabend, *Against Method*, NLB (1975), I. Lakatos, "Falsificationism and the Methodology of Research Programmes," in I.

Lakatos and A. Musgrave, *Criticism and the Growth of Knowledge*, Cambridge University Press (1970).
16. *Eye and Brain*, McGraw-Hill (1966), p. 225.
17. Of course, the conception of science implicit here cannot be too narrow. For details see Chs. 3-5 of my *Radical Knowledge*, Hackett (1981).
18. The rather common failure to understand this point is behind the claim that evolutionary theory is tautologous.
19. For example, J. Bronowski in his television series (and book), 'The Ascent of Man.'
20. *Radical Knowledge, op. cit.*
21. *op. cit.*, especially pp. 228-235.
22. *op. cit.*, pp. 242-244.
23. *op. cit.*, p. 228.

7 Towards a Future Epistemology of Science

1. The Problem

The latter part of this century has been a time of upheaval for the philosophy of science. It is tempting to describe the situation as a struggle between two schools of thought: the logical (also called analytic-empiricist) and the historical (also called socio-historical). The first school of thought has inherited the house of philosophy that the Logical empiricists and the Popperians built, while the second is often envisioned as a wolf (either Kuhn or Feyerabend) threatening to blow down the entire structure. It is felt in some quarters that the wolf has already won and that if there still seems to be a struggle it is due only to the inertia of a long entrenched analytic reflexes. And so we hear Richard Rorty, for example, proclaim that the age of philosophy as epistemology, i.e., as the discipline that spells out the justification of scientific practice, has come to an end and should be replaced by the age of hermeneutics, i.e., the attempt to bring about empathy toward different approaches to nature and man's place in it.[1]

I would like to think, however, that there are some different, valuable lessons to be gained from the contemporary upheaval, and that in drawing them we may realize the directions that a new epistemology of science ought to take.

First of all, it seems to me that the division of the field into the two schools of thought, even though warranted at first sight, falls apart under closer scrutiny. Why this is so is rather important and revealing, as we shall see.

What I have called the 'logical' school of thought would hold that, as Carl Hempel says,

> ... it is specifically the task ... of the philosophy of science to exhibit, by means of 'logical analysis' or 'rational reconstruction', the logical structure and the rationale of scientific inquiry. The methodology of science, thus understood, is concerned solely with certain logical and systematic aspects of science which form the basis of its soundness and rationality — in abstraction from, and indeed to the exclusion of, the psychological and historical facets of science as a social enterprise.[2]

The principles established by the methodology of science can then serve as 'conditions for the rational pursuit of empirical inquiry, as criteria of rationality for the formulation, test, and change of scientific knowledge claims'.[3]

The historical school of thought, on the other hand, 'rejects the idea of methodological principles arrived at by purely philosophical analysis'.[4] Instead it presumably insists that an adequate account of science must be consistent with the actual practice of science, and of course that practice can be found instantiated in the history of science. The claim made by the 'historical' side against the 'logical' side is that the crucial methodological principles do not agree with the way science has been practiced. This may not seem like much of an objection, for the proponents of scientific methodology do not claim that they are presenting an account of how scientific thought actually proceeds but of how it ought to proceed. Indeed, philosophy of science not only discovers but also sets itself, in Lakatos' words, 'as a watchdog of scientific standards'.[5]

Nonetheless the argument from scientific practice can be decisive if one simply asks why should any scientist wish to follow the norms produced by philosophical analysis. The reason is presumably that it pays for him to do so, that is, he increases his chances of scientific success by acting in accordance with the scientific method. Scientific rationality is rationality at all because it is behavior that accords with those procedures that secure knowledge about the world. What philosophical analysis does, then, is find out those procedures; it abstracts from the practice of science those aspects that make for so keen a probe of nature. Thus philosophical analysis, for all its aprioristic airs, is as empirical about science as science is about the world. This point is acknowledged by Hempel, who recalls that 'the efforts of analytic empiricists to "explicate" norms for scientific inquiry, conditions of empirical significance, criteria of demarcation for scientific hypotheses, rules for the introduction of theoretical terms, and the like, were never undertaken in a purely *a priori* manner. Explications were constructed with an eye on the practices and the needs of empirical science'.[6] It is precisely the pressure of scientific practice that led to the eventual rejection of the verifiability criteria, for example. 'Thus, explication in the sense of analytic empiricism', Hempel adds, 'has been guided to a considerable extent by close attention to salient features of actual scientific procedures and the logical means required to do justice to them. This process of rational reconstruction, as conceived especially by Carnap and some like-minded thinkers, does, it is true, lead to idealized and schematic models; but these are formulated in consideration of the kinds of scientific systems and procedures whose rationale they are intended to exhibit'.[7] More recent practitioners or the art of rational reconstruction would have to make similar admissions. Thus Lakatos, who argues for what he calls 'statute law' about science says that 'Demarcationists reconstruct *universal* criteria which explain the appraisals which great scientists have made of *particular* theories or research programmes'.[8]

This explains the proper relationship between philosophical analysis (whether it be called 'logical', 'linguistic', or something else) and methodological norms. As I will argue below, in cases such as that of Lakatos, the philosopher should admit that he is drawing his methodology from some specific episodes in the history of science, which for some reason he finds particularly compelling. But in other cases, and characteristically in analytic philosophy of science, the philosopher has internalized a certain conception of science (often a collection of myths about science) and then by 'analysis' discovers what he already 'knows' about science. It is not surprising that the 'concepts', which he tries to articulate and clarify, have certain 'logical' properties and bear certain 'logical' relations to other such 'concepts'.[9] But since he brings to his investigations the logical machinery of the axiomatic approach, and since he is not aware of having obtained his initial conception of science directly from experience, he is doubly impressed with the *a priori* character of his results. This is a common experience in any field of endeavor: the more familiar something is, the more it becomes part of us as we grow with it, the more difficult it is to see how it could be otherwise. Unfortunately since the original conception of science by the very nature of its acquisition is not the result of close scrutiny,[10] once the methodology born out of philosophical analysis is put to work it gets into grave difficulties. There ensue then a series of 'logical' arguments to patch things up. But this is just the way in which somewhat more realistic ideas about the practice of science manage to make their way into the arena of 'pure' conceptual philosophy. Eventually it is shown, as Kuhn and Feyerabend have, that such methodology would lead science nowhere. More to the point, it is shown that the scientific method was not only violated in significant episodes in the history of science (Galileo's case for Copernicanism, Einstein's theory of relativity, etc.) but furthermore that it had to be violated for 'progress' or 'success' to result.[11] The situation is simple then. Science is very successful. Rationality (read methodology) was supposed to give us the key to scientific success. But it actually gets in the way of such success. The choice is clear: either success or rationality. But rationality made sense only with respect to success.[12]

The analytic establishment was predictably aghast at a view that presented science as an irrational enterprise, i.e., as an enterprise that did not agree with any of the favorite 'aprioristic' conceptions of rationality. But then, if my thoughts on the matter are correct, those very conceptions of rationality are so ill-founded that we need not mourn their passing. Thus even though the contemporary controversy is often stated in terms of a dispute as to whether science is a rational activity or not, a deeper look shows instead a transformation of the problematic of the field. Perhaps Kuhn is correct in saying that 'if history or any other empirical discipline leads us to believe that the development of science depends essentially on behavior that we have previously thought to be irrational, then we should conclude not that science is irrational, but that our notion of rationality needs adjustment here and there'.[13] Or perhaps, as Feyerabend seems to suggest, rationality is nothing but one of those grandiose verbal banners that command assent from all respectable citizens but which turn out to be irrelevant to the practice of science except for their propagandistic value. I will reserve judgement on this issue until late in the paper. Incidentally, it must be realized that the 'problem' presented by the historical school is not merely a revival of skepticism. In the case of skepticism we 'know' that the conclusions of science are on the whole warranted, we feel it in our bones, but we cannot prove it. In the present case, by contrast, we find out that the procedures for success so in agreement with our peculiar ossification actually obstruct the results they were supposed to facilitate.

The logical approach turns out to be a poorly realized form of the socio-historical approach then. This discovery is not a small matter, for in spite of Hempel's acknowledgement that the logical empiricists had the actual practice of science in the back or their minds, the dominant philosophical movement in this century has been decidedly logistic, one might say axiomatic, or to use Lakatos' phrase, 'neo-Euclidean'. But if the considerations of this paper are correct, this entire edifice of sharp logical distinctions has been built on philosophical quicksand. How could this come about? It is not the purpose of this paper to explain the quirks of the history of philosophy, but perhaps a few speculations are in order. It was not so much lack of perceptiveness that dragged philosophers down the blind alley of analytic philosophy. They were swept along rather by the overwhelming intellectual need, as it was felt in the later part of the 19th century, to demonstrate the correctness of mathematics. With the discovery of non-Euclidean geometries, mathematics entered what might be called a period of crisis. Until then it had been thought that Euclidean geometry not only rested on unshakable foundations but that it held a promissory note for the eventual demonstration of the rest of mathematics. With the questioning of Euclidean geometry the promissory note was no longer available and there was urgency to demonstrate the truth (in some sense) or at the very least the consistency of mathematics. The subsequent work on the foundations of mathematics suggested very strongly that either logic or some other appropriate axiomatic method would be the answer to the purists' hopes. With methods of such extraordinary precision, as they were thought to be, one could finally prove what was what. The current was so irresistible that some of the most important figures involved (e.g., Frege and especially Russell) thought that here was a method that would set things right in philosophy as well: no more flights of fancy, no more taking things for granted. Thus an attempt to preserve mathematics' virtue became the standard of philosophical good sense, and remained so even long after Gödel showed that not even the consistency of mathematics could be proved.

Since aprioristic approaches by their very nature exclude socio-historical conditions, it is not surprising that philosophers found all such irrelevant, even though, as Hempel now recognizes, as philosophers went along in their merry formal ways they could not help keeping an eye on the actual practice of science. And the day finally came in which the very pure philosophical analysis was found to have been infected all along with all those socio-historical considerations that its forefathers thought a good logical vaccine would vanish from the world of philosophy for ever. This brings us back to the contemporary scene.

It may be thought that the mere failure *so far* to provide an *a priori* conception of science that fits the practice of science (within reasonable boundaries) does not constitute an argument against the possibility of such a conception. But we must realize that as soon as the considerations from the practice of science are seen as legitimate, the cat is out of the bag, *even if Kuhn's and Feyerabend's interpretations of the history of science turn out to be mistaken*. For what is compromised by these considerations is the very purity of philosophical analysis. We simply are confronted with a situation in which conceptual ('logical', 'linguistic', etc.) investigations are to be settled by 'dirty' empirical findings, in which conceptual conclusions must be *justified* by such findings. An attack on Kuhn and Feyerabend's interpretations of history may perhaps defend a particular methodology, qua methodology, but its mere presence advertises the loss of virtue of the aprioristic movement in philosophy.

Lakatos is not deterred, however. He argues that for considerations about the practice of science to be effective against any one methodology they must first be about the correct practice. Thus, for example, one should not criticize Carnap because his logical models do not fit astrology. This means that we must first identify correctly what counts as science (otherwise all those sociological and historical investigations go for naught). But how can we do this unless we have a prior system for demarcating science form other activities?[14]

Of course, there are those who think that it is not for them to settle what science is before they investigate it. They start by looking at what is called science and try to discover features that may hold the key to our understanding it. This move meets with derision on Lakatos' part. Kuhn, who does this, and Polanyi before him, and now Toulmin with his evolutionary epistemology, are all forced, Lakatos claims, to take the word of a scientific elite, for all of them insist that only those who have undergone the training of science are in a position to realize the appropriateness of any particular decision in science. This elitism does not permit the imposition of any external rational guidelines (Lakatos is all for letting the layman serve as Juror in disputes about the rationality of scientific practice, a task facilitated by Lakatos' statute law — that anyone can apply — in contrast to the case law of the scientific elitists which only they can apply properly for only they have developed the appropriate instincts). But surely, no matter how ugly the charge of elitism might be, the position that favors it cannot be considered wrong merely on aesthetic grounds (not from a rationalist point of view, at any rate). The main problem with elitism, Lakatos believes, is that it cannot possibly explain how the changes in science were really appropriate. If this elite is not acting according to rules that can be recognized independently as rational, if their proceedings are guided, if at all, by some sort of community subconscious instead, then whatever success they achieve must be ascribed to the Cunning of Reason, or some other equally ludicrous Hegelian mechanism.

Such is the price to be paid for insisting that some scientific views are better than others while following a 'naturalistic' account of science (as Feyerabend calls the socio-historical approach). But what alternative does Lakatos offer to this elitist naturalism? He is weary of absolute demarcation criteria or methodology, for he is very well aware of the downfall of his demarcationist predecessors. Indeed he calls himself 'a fallibalist with regard to demarcation criteria, just as I am a fallibalist with regard to scientific theories'.[15] Not only does he admit that methodologies can be criticized by use of the historical record, he himself proposes most ingenious historiographical criteria by which to judge the comparative merits of different methodologies.[16] Of course one should not expect every methodology to fit the practice of science perfectly. For one thing some science is better than other. But even in cases of excellent science, the developments are not perfect. A superior methodology, when applied to such episodes, will permit us to understand what elements made for the success of that science, and presumably which were not essential to that success.[17]

This approach sounds quite sensible, but there are two important points to consider. The first one is that we no longer are in the realm of *a priori* philosophy. Such philosophy seems to be reduced to the role of providing a starting point for our investigations about the nature of scientific knowledge. But the end point is dictated at least in great part by our experiences along the way (this result is on a par with claiming that mathematics is justified only insofar as it serves the ends of physics well, and not because of consistency, completeness or some other *a priori* criteria).[18] And second, this mixing of philosophical approaches, this compromise suggested by Lakatos, looks very suspicious. Did he abstract his criteria from the workings of the great scientists? Or did his armchair travels through the Third World just seem conveniently at an end precisely when his ideas offered a close enough fit to what the great scientists had done? In the first case, how did he know that such were the right episodes from which to abstract his criteria? In the second case, how did he know that was the appropriate place to stop? In both cases, why these episodes (the great scientists' struggles) and only these? It seems that Lakatos is guilty of the same charges he leveled against Kuhn, Polanyi, Toulmin and all the elitist naturalists. He has simply taken what was already recognized as great science and distinguished patterns in it to which he now intends to have the rest of science conform. In other words, he has accepted the judgement of the scientific elite about the history of science. That is exactly what he accuses the elitists of doing. The only difference is that the *other* elitists did not wrap their investigations with the magical mantle of Reason, nor made mysterious appeals to an even more mysterious Third World of propositional knowledge. Another difference is that the others generally impose fewer restrictions on the nature of the mechanisms that may account for the success of any particular episode in the history of science, thus they allow sociology and anthropology roles that had been hitherto reserved to logic alone.[19]

We began by characterizing the present situation in philosophy of science as a struggle between two diametrically opposed schools of thought. But we have seen that both sides, as well as the suggested compromises, differed only to the degree to which they pretended to be something other than naturalism. In the long run they must all hold that, in Feyerabend's words, 'Reason receives both its content and its authority from practice. It describes the way in which practice works and formulates its underlying principles.'[20] We may still argue about statute vs. case law of rationality (or indeed about no law at all), but it seems that naturalists we must all be. Some may think that this result is of the greatest significance, for it points out precisely the way in which the dominant movement during much of this century drove philosophy into a blind alley. Perhaps in naturalism we now have an approach, or a series of approaches, divested of pretensions to a knowledge not tied to this world, and capable of making philosophy a relevant discipline once again.

I think, however, that naturalism has great defects of its own, whether it offers a methodological face or not. The logical approach was blatantly ahistorical but could not escape history. The pitfall of naturalism, I will argue, is precisely that it embraces history. It seems to me that an appropriate view of science will have to take into account that science is an activity which by its very nature transcends history. I mean nothing mysterious by such remark. Science not only constantly transforms its own content, but in the process of that transformation it also changes the range of environments to which it applies, and sometimes those environments themselves. Science may always have a foot in its history, but it always faces a universe unknown. A proper account of science must go beyond the present and past scientific practice; it must concern itself with the future as well. This is not to say that the philosopher of science must now become a futurist, let alone a divine of some sort.

Let me illustrate what I have in mind by means of an analogy. A man is mysteriously transported to a planet that seems totally unrecognizable to him. He does not know whether he will ever be able to go back, but feels that in order to take proper stock of his situation he must get to know as much about this planet as possible. His determination requires that he travel all over the planet. This planet has a great variety of environments, however, all offering new opportunities as well as unusual dangers. The means of transportation that the man chooses must fit the particular environment. The canoe that allows him to travel through the swamps will not do in a sea, let alone a desert. There is a reward in being able to use the same materials as much as possible. But an insistence in doing so may prove too rigid an approach, with fatal consequences. Analogously, our universe has changed the aspects that it has presented to us in so radical a manner that we might have just as well traveled through a great variety of environments, with science as our means of transportation. Different ideas definitely looked more appropriate than others in certain circumstances. And it seems that the same situation will continue, as I will argue later. But shouldn't we expect also that some methodological standards will be more appropriate than others as well? Standards, as Feyerabend says, "are intellectual measuring instruments; they give us readings not of temperature, or of weight, but of the properties of complex sections of the historical process."[21] To assume, as Lakatos does, that from the scientific practice already behind us we can tell what standards are appropriate to all situations, including those that we cannot even fathom yet, is analogous to assuming that we can satisfactorily answer 'the question what measuring instruments will help us to explore an as yet unspecified region of the universe. We don't know the region, we cannot say what will work in it ... (as an example consider the question how to measure the temperature in the center of the sun, put at about 1820)'[22] Similarly with standards, then, 'Are we supposed to know them even before these sections [of the historical process] have been presented in detail? Or is it assumed that history, and especially the history of ideas is more uniform than the material part of the universe? The man is more limited that the rest of nature?'[23]

The problem with naturalism, then, is that it learns too well a lesson that history doesn't really teach. It takes limited experience and turns it into eternal expectation. Feyerabend, who is not a naturalist, his predilection for history notwithstanding, realizes this point: 'Having chosen a popular and successful practice the naturalist has the advantage of "being on the right side" at least for the time being. But a practice may deteriorate, or it may be popular for the wrong reasons ... Basing standards on a practice and leaving it at that may for ever perpetuate the shortcomings of this practice'.[24] As for those who admit that different circumstances may call for different standards, but trust the elite to make the right decision every time, we may point out with Lakatos that 'surely degeneration is at least possible even within a "scientific" community. After all, a victory of Lysenko's research programme might have been brought about in the West too if all Lysenko's opponents had died of natural causes within a couple of months, instead of being sent to concentration camps and killed. Would Lysenko's theory then have been vindicated through having been produced by a Mertonian or Polanyiite scientific community? Obviously not. Might is not bound to be right even within a perfectly "rational" scientific community. One may obviously even have perfect consensus and degeneration at the same time'.[25]

2. The Solution

How can such an epistemology be constructed? Perhaps it is only fair to indicate the direction I have followed elsewhere.[26] The following sketch, however, is not intended as a comprehensive summary of positions previously developed, but simply as an example of a way in which the requirements I have envisioned for a future epistemology may be met.

It pays to recall, I think, that science is a product of intelligence and that intelligence is an instrument of adaptation and itself the result of evolution. What distinguishes intelligence from other mechanisms of interaction with the world associated with the central nervous system is that intelligence transcends the ability to respond merely to the immediate exigencies of time and space.[27] It is this freedom of response that permits intelligence to form sweeping views of the world and the means for criticizing them. But science does more than extend the functions of intelligence. Kuhn and Feyerabend, as well as Popper long before them, pointed out that scientific views constitute the spectacles through which we view the world. That is, science is the very means by which intelligence deals with the world in large.[28] Furthermore, science, like other 'telescoped' functions of intelligence, is a communal enterprise that involves division of labor and is carried out in a social milieu. It pays, therefore, to study the ways in which this social character may shed light on some of the questions that confront the epistemology of science.

It should come as no surprise that this sort of epistemological account puts a high premium on evolutionary considerations. For our present concerns the most important result is the suggestion that scientific knowledge has adaptive value. Suggestions of this sort have been remarkably unsuccessful before, in part because the biology they assumed was completely implausible (in the case of Ernst Mach, for example, he tied every scientific idea to the mechanisms of evolution, which would demand biological, i.e., genetic, change at the same pace of scientific change), and in part because of misunderstandings and prejudices masquerading as solid objections (for example that scientific theories are seldom — if ever — accepted or rejected on their adaptive value, that in fact scientists who make such decisions may not even be aware of what practical consequences those theories may have, that they simply accept those views that best satisfy their intellectual curiosity. The reply is that, as we saw in Chapter 6, most scientific work does not deal directly with nature — a lot of it aims to articulate theory or experimental technique, or to integrate phenomena into the causal network for the first time. All this work may be seen as an indirect means of dealing with the world, and as part of a whole that has adaptive value. As for curiosity, it can be shown, even in the case of animals, that it also has adaptive value).

To say that science has adaptive value amounts to saying that science enables us to do the following things:

(1) deal with greater ease with our environment (our 'niche');
(2) increase the number and diversity of environments that we can deal with (enlarging the 'niche');
(3) cope with a continuously changing environment (which puts a premium on flexibility of response).

The direct and indirect ways of connecting with the world to which I alluded above are normally employed to achieve aim (1) and thus are designed and perfected to deal with a specific environment. But then the satisfaction of aims (2) and (3) requires that science be so organized that alternatives to the dominant views of the time be routinely generated and developed. This could enable science to change with a changing environment and so on (similar to the manner in which genetic variation would facilitate the adaptation of species).[29] This production and articulation of alternatives is the key to correcting the excesses of naturalism, for, as Feyerabend says, just as 'the inadequacy of standards becomes clear from the barrenness of the practice they engender, the shortcomings of practices often are very obvious when practices based on different standards flourish'.[30] The very social structure of science must preclude, then, tyranny by the elite, for *epistemological* reasons, i.e., for reasons connected with the success of the enterprise. This is in perfect accord with Feyerabend's point that, even if one thought, as rabid empiricists are prone to think, 'the only pressing reason for changing a theory is disagreement with the facts' (which leads them to discard the discussion of alternative views of the world), it must still be realized that 'not only is the description of every single fact dependent on some theory ... but there also exist facts which cannot be unearthed except with the help of alternatives to the theory to be tested, and which become unavailable as soon as such alternatives are excluded'.[31]

If science has adaptive value, as I envision, it must have a structure consistent with the achievement of such value. And that structure requires as its foundation what we might call 'the principle of intellectual freedom'. But then not only do we have a check on elitism, this evolutionary account of science also suggests strongly that the matter of the rationality of science is to be determined not by the adherence (or lack of it) by scientists to particular methods or standards, but by whether science has the structure that permits it to fulfill its function. On the whole it seems to me that science has such a structure, and not in spite of but precisely because of the sorts of historical analysis provided by Feyerabend.[32] His attacks on rationality (at least to date) have not been directed against the social conception, but against a rationality that depends on the behavior of individual scientists, and that resembles in character the standard aprioristic methodologies.[33]

I trust that this sketch indicates that developing an appropriate new epistemology is by no means hopeless. The events of the latter part of the century, far from bringing philosophy to an end, have freed it from its logical shackles, and have as a result opened up a host of new opportunities. The field may have never been as exciting!

It would be felicitous to end on this uplifting note, but I find it wise to touch briefly on three matters of some relevance to the considerations advanced in this paper. The first one is Hempel's claim that 'rationality seems ... intelligibly attributable only to behavior that is causally traceable to reasoning or deliberation about suitable means for attaining specified ends'.[34] If Hempel is correct, rationality cannot be understood in the social terms that I have suggested. The benefits that accrue to the scientific enterprise if it is structured as I suggest are not, as Hempel points out in making a similar objection to Kuhn, 'objectives pursued by the reasoned adoption of a group procedure that has been deliberately chosen as an optimal means to the end of achieving those beneficial effects'.[35] If so, science as I have described it and recommended it 'would have to be viewed as akin to certain other social institutions or behavior patterns which in anthropology and sociology are said to be "latently functional" on the ground that they fulfill certain requirements for the survival or the "success" of the group concerned, without, however, having been adopted by deliberate social choice as a means to that end. Now, such mode of behavior might be called *adaptive*, but surely not *rational*: they are not adopted as result of goal-directed reasoning.'[36]

We must realize, however, that no philosopher of science, including the most staunch methodologists, would pretend that his view of science was *consciously* acknowledged by most if not all scientists. There would have never been any point to the work of methodologists if they simply summarized what scientists say, instead of putting forth what scientists *ought to do* and in fact *do* when practicing good science. The universal criteria that Lakatos has in mind, for example, is that which, according to him, 'great scientists have applied sub- or semi-consciously'.[37] Awareness of the methodology was never a requirement for rationality. Otherwise much philosophy of science would have been out of business from the start. But then 'deliberate social choice' is even less reasonable as a requirement for the ascription of rationality. In the perhaps parallel case of the rationality of an agent's beliefs, a corresponding requirement would make such rationality hinge on the agent's deliberations. As Wittgenstein pointed out, however, one may believe that X without saying to oneself (not even ever so softly) that X. In that case it would be strange to insist that appropriate deliberation must be a criterion of rationality in addition to the appropriateness of the means-ends relations concerning the beliefs in question.

Another objection, this one against my stratagem to develop a new epistemology, would come from Rorty's emphasis against accounts of knowledge as discriminative behavior. That is the sort of account that should be expected from a position that begins by tying science to intelligence in an evolutionary context. Such accounts, Rorty would hold, are not really of knowledge but of simply 'reliable signaling', of the sort of thing 'manifested by rats and amoebas and computers'[38] but presumably not by beings capable of the type of awareness that Sellars would place in the 'logical space of reasons, of justifying and being able to justify what one says'.[39] Only in this second sense can one hope to have knowledge, whereas in my sense we are dealing merely with the 'ability to respond to stimuli'.[40] Rorty follows Sellars in claiming that 'such ability is a causal condition for knowledge but not a ground for knowledge',[41] for, as he also says, 'there is no such thing as a justified belief which is nonpropositional, and no such thing as justification which is not a relation between propositions'.[42]

I can see no justification for such linguistic ascent, however. The push-pull notion of discriminative behavior that Rorty offers is not an accurate rendering of the properties of intelligence described above. Indeed intelligence transcends the mere 'ability to respond to stimuli' without being beyond the reach of rats and other animals that may not give a hoot for Sellars' 'logical space of reasons'. Of course it is the degree of complexity and sophistication of our intelligence that permits us to develop scientific knowledge. But the proof of this pudding is, I have argued elsewhere, in the enhanced ability to deal with the world.[43] This test is similar to that which a computer, or an alien being (or in a science-fiction scenario, a being that seems very human but of whom we need to feel very sure) must pass for us to consider an ascription of intelligence. In capsule form, the test in question is just a generalized Turing test: place the subject in an open environment and see how well it can manage. When the social undertaking that we call science passes the corresponding test we say that we have acquired scientific knowledge. Whether language (and the allegedly concomitant 'logical space of reasons') is indispensable equipment to pass this test does nothing to demonstrate that we are no longer in the realm of 'discriminative behavior'.[44] This may be 'reliable signaling', but it still earns the title of knowledge. In any event I do not see why it may not.

I will close by making clear that I do not suggest a mechanism for arriving at the *right* procedure in every occasion. The switch to the social conception of scientific rationality is at least in part an attempt to think about the development of science in different terms. What I suggest instead is a mechanism (if one may call it that) to increase our chances of taking advantage of theoretical and experimental opportunities. The pressure of competing practices serves to keep the leading practice on its toes, for example, but it also provides for new ways of looking at the world which in a ripe moment may prove invaluable. This is not to say that I am claiming title to a recipe for achieving 'objectivity' in the old-fashioned, empiricist way. All sorts of ideas may be of great service as we interact with the universe, without being *true* or *right* in the sense of telling us what the world is *really* like. Nor do I suggest a way of progressing toward any such ideal. Indeed, I would argue to the contrary that a proper evolutionary account, far from guaranteeing such objectivity, will usher in a thoroughgoing relativism.[45] (See Part I, especially Chapter 1) If performance is to serve as the model of scientific understanding, neither ontology nor logical propriety (e.g. consistency)[46] can be used as a measure of objectivity. This result should not be any more frustrating to the aims of science than diversity of adaptation should be to life. Quite the opposite, in the opportunism of life we may see a good example of how to prepare for the vicissitudes of the future.

Notes

1. Richard Rorty, *Philosophy and the Mirror of Nature*, Princeton University Press, (1979).
2. Carl G. Hempel, 'Scientific Rationality: Normative vs. Descriptive Construals', in *Wittgenstein, the Vienna Circle and Critical Rationalism* Proceedings of the 3rd International Wittgenstein Symposium, (1978) p. 291.
3. *Ibid.*
4. *Ibid.*, p. 292.
5. Imre Lakatos, 'Understanding Toulmin', reprinted in his *Mathematics, Science and Epistemology*, J. Worrall and G. Currie (eds.), Cambridge University Press, (1978), p. 226.
6. Hempel, op. cit., pp. 298-99. The point was well made by John Dewey at the time. See his *Logic: The Theory of Inquiry*, p. 9.
7. *Ibid.*, p. 299.
8. Lakatos, *op. cit.*
9. The situation would not be all that different if he assumed a different conception of science, though he would talk about other 'concepts' and other 'logical relations'.
10. A rather hopeless task for a view that frowns on history.
11. Today's intellectual wisdom about the 'empiricist' Galileo, for example, is typified by claims such as Rorty's that the empiricists 'were doubtless also right in commending Galileo for preferring his eye to his Aristotle'. (*op. cit.* p. 246). Standing this wisdom on its head, Galileo openly admitted that Copernicanism was a view born refuted: 'There is no limit to my astonishment, when I reflect that Aristarchus and Copernicus were able to make reason so to conquer sense that, in defiance of the latter, the former became mistress of their belief'. (Galileo Galilee, *Dialogue Concerning the Two Chief World Systems*, Berkeley (1953). p. 328). Indeed, Copernicus 'with reason as his guide ... resolutely continued to affirm what sensible experience seemed to contradict'. (*Ibid.* p. 335). Theory can, then, overturn the verdict of experience by introducing an alternative body of experience more suitable to one's views. Sometimes this is done, as in Galileo's reply to the Tower Argument, by offering a different set of what Feyerabend calls 'natural interpretations' (thus the stone dropped from a tower does not really move straight down, it only *seems* to). Other times the testimony of the senses is overruled by 'the existence of a superior and better sense than natural and common sense' (the telescope), which joins 'forces with reason'. The eyes, whose testimony Galileo fortunately disbelieved, actually refuted the Copernican view, for the magnitudes of the planets were far less than they should have been. However this is done, the result is devastating to methodologies which have as a cornerstone the claim that experience determines the worth of theory and not the other way around. To rub salt on the wounds, it also seems that the defense of cherished methodological principles (agreement with the facts, suspicion of ad hoc stratagems such as Galileo's circular inertia, and so on) was left to the Aristotelians.

12. The structure of the argument is simply that of a *reductio*. It is truly amazing that many critics have chided Feyerabend for 'using reason to undermine the authority of reason'.
13. Thomas Kuhn, 'Notes on Lakatos', in R.S. Cohen and R. C. Buck (eds.), PSA 1970, *Boston Studies in the Philosophy of Science*, vol. VIII (1971), p. 144.
14. Lakatos, *op. cit.*, pp 235-253.
15. *Ibid.*, p. 242.
16. Lakatos, 'History of Science and Its Rational Reconstructions', reprinted in his *The Methodology of Scientific Research Programmes*, J. Worrall and G. Currie (eds.), Cambridge University Press, (1978), pp. 102-138.
17. Lakatos' methodology of research programmes can manage many of the problems that plagued the standard methodologies, but only at the price of abandoning the right to advise the scientist. His methodology, as he says, *'allows people to do their own thing but only as long as they publicly admit what the score is between them and their rivals'*. There is freedom ('anarchy' if Feyerabend prefers the word) in creation and over which programme to work on but the products have to be judged. *Appraisal* does not imply advice'. ('The Problem of Appraising Scientific Theories', *Mathematics, Science and Epistemology, op. cit.,* p. 110). This very tame notion of rationality seems more relevant to the historian of ideas than to the scientist.
18. For a presentation of this sort of position in mathematics see Morris Kline's *Mathematics: The Loss of Certainty*, Oxford University Press, (1980).
19. Some of the other arguments that Lakatos advances against Toulmin's views are uncannily similar to those Feyerabend used against Lakatos' own methodology of research programmes (e.g. the claim that since Toulmin is forced to employ the Long Run his view has no teeth).
20. Paul Feyerabend, *Science in a Free Society*, NLB (1978), p. 24.
21. *Ibid.*, p. 37.
22. *Ibid.*
23. *Ibid.*
24. *Ibid.*, p. 25.
25. Lakatos, 'The Problem of Appraising Scientific Theories', op. cit., p. 115.
26. Particularly in my *Radical Knowledge*, Avery Publishing Co. and Hackett Publishing Co., (1981).
27. *Ibid.*, Chapter 3.
28. *Ibid.*
29. The environment itself may change, or by our changing our views we deal with it in different terms, which for the present purposes amounts to the same thing. This situation is likely to continue, at least for the foreseeable future. As an example consider that space science, which is in its infancy, permits us to make astronomical observations that sidestep many of the difficulties that have limited earth-bound astronomy (atmospheric absorption and distortion, columns of air in the telescope, gravitational bending of large lenses, etc.). From the technical point of view this may be as advantageous as the telescope was over the eye — and without as many epistemological problems.
30. Feyerabend, *op. cit.*, p. 25.
31. Paul Feyerabend, *Against Method*, NLB, (1975), p. 39.
32. *Radical Knowledge*, op. cit., Chapter 4.

33. In his *Science in a Free Society*, Feyerabend argues that even if it could be shown that science is reason, it would still be necessary to show that reason is better than other human activities. For my comments see Appendix A.
34. Hempel, *op. cit.*, p. 299.
35. *Ibid.*, p. 300.
36. *Ibid.*
37. Lakatos, 'The Problem of Appraising Scientific Theories', *op. cit.*, p. 110.
38. Rorty, *op. cit.*, p. 182.
39. *Ibid.*
40. *Ibid.*, p. 182.
41. *Ibid.*
42. *Ibid.*
43. *Radical Knowledge*, op. cit., Chapters 4-5.
44. Surely, this is more than 'mere' reliable signaling or discriminative behavior. But making a special category for it on the grounds of degree of complexity will not help Rorty much. In that case he would be guilty of offering a false dilemma.
45. *Radical Knowledge, op. cit.*, Chapter 2-3.
46. See my 'Allowing Contradictions in Science', *Metaphilosophy*, vol. 13 no. 1 (January 1982), p. 75.

8 Hull, Biology, and Epistemology

For years I have enjoyed David Hull's keen eye for biological detail and his use of that detail to illuminate philosophical problems. In his "A Mechanism and its Metaphysics: An Evolutionary Account of the Social and Conceptual Development of Science," Hull (1988) displays that keen eye once more, but I think that this time he falls short of his ambitious philosophical goals.

As Hull points out, the majority of evolutionary epistemologies have tried to explain conceptual development in science by devising analogies to biological evolution. Unfortunately the selection mechanisms so far proposed have not been satisfactory. Hull suggests that the solution lies in a more general analysis of selection that would include both biological and conceptual evolution, and he thinks that he has just the right mechanism to buttress that analysis. I am afraid, however, that his analysis does not work.

On a second front, Hull's view of science encounters an obstacle that hinders so many other naturalistic epistemologies: the jump from a description of how science works to epistemological prescription. Even quasi-naturalistic approaches, such as that of analogical evolutionary epistemology crash against this obstacle. The reason is that even if a tight analogy to biological evolution could be established, we could still say nothing about the rationality of science and about other normative issues typical of genuine epistemology (for surely, since biological evolution is not rational, the fact alone that scientific evolution is just like it is not a good reason for claiming that science is rational). Some strands in Hull's view could help it surmount this obstacle, but Hull does not take them very far. I will deal with each of these two main objections in turn.

For selection to exist, Hull says, we must have appropriate interactions between what he calls "replicators" and "interactors." A replicator is "an entity that passes on its structure largely intact in successive replications" (p.134). An interactor is "an entity that interacts as a cohesive whole with its environment in such a way that this interaction causes replication to be differential" (p.140). In science the replicators are scientific ideas, including beliefs about the goals of science, the proper ways to realize those goals, problems and their solutions, modes of representation, data reports, and so on (p.140). I take it that ideas about the structure of the world would be included, since they clearly are "elements of the substantive content of science." The interactors are chiefly the scientists themselves (p.140). These conceptual replicators are on a par, presumably, with hitherto standard replicators such as genes. "Conceptual replicators," says Hull (p.140), "interact with that portion of the world to which they ostensibly refer no more directly than do genes with their more inclusive environments. Instead they interact only indirectly by means of scientists." Now, since conceptual replicators interact with the world, albeit indirectly, we might think that Hull is going to propose that (some) scientific ideas have adaptive value, as (some) genes do, and that this value is the key to his evolutionary epistemology. But Hull does not develop this line of thought, and apparently to the contrary, insists later that science works best when scientists pay attention to those problems they find most interesting rather than to "those that are currently most relevant to our survival" (p.154).

Without such a development, Hull's analysis of selection becomes very perplexing. According to him "selection is a process in which the differential extinction and proliferation of interactors *cause* the differential perpetuation of the replicators that produced them" (p.134). If his definition is correct, scientific change is not a selection process. There are two main reasons why. First, the interactors in question are human beings, scientists. It is difficult to grasp what Hull could mean here by the differential extinction and proliferation of scientists, unless science is a far nastier business than even its most unperceptive critics make it out to be. If Hull claimed that certain scientific beliefs had survival value then we could see why scientists who held those beliefs would be favored in differential extinction and proliferation. As it stands, however, this view is too crude, and Hull shies from it anyway.

Perhaps the differential proliferation in question is of the scientists *qua* scientists. The successful scientist is just the one who gets his ideas accepted. Interaction and replication would then be defined in the context of a social environment of scientific rewards and punishments. The criterion of success for this social interaction would be success in social replication. In one reading of this view, the difference between interaction and replication would collapse. This collapse, however, cannot be permitted by Hull's clear separation between the two, since conceptual replicators interact only indirectly with the natural world by means of scientists (who presumably interact directly). Other less troublesome "social" interpretations run into trouble, nonetheless, because Hull does require interaction (albeit indirect) between the replicators and "that portion of the world to which they ostensibly refer" (p.140). Once we accept this point, a purely social criterion for success is no longer acceptable.

Whatever the prospects for this first clause of Hull's definition of selection, I do not see how the second clause can be salvaged. For that clause leads us to infer that scientific ideas and beliefs (the replicators) must have *produced* the scientists (the interactors). In the case of biological evolution this analysis of selection makes sense: genes do produce more inclusive phenotypes; those phenotypes interact with the environment; and the successful phenotypes pass on the genes that produced them. "Produced" here clearly indicates a causal relation, and since Hull wants to be taken literally, not analogically, his extension of selection to scientific change becomes absurd.

As far as the second charge is concerned, i.e., that his view does not amount to a genuine epistemology, Hull cannot say that his aim is only to explain scientific change, since he seems to believe that his account lays bare the rational character of science. Thus for example, he argues that this socio-psychological mechanism explains the existence of priority disputes within and between groups, and that all such disputes are of "rational interest" (p.134). Indeed he finds that this is one of the advantages of his account over accounts such as Lakatos' in which only disputes between research programs can be of rational interest while those between scientists promoting the same program result merely from "vanity and greed for fame" (p.139). Hull's mechanism is that scientists strive to have their ideas accepted by other scientists ("conceptual inclusive fitness") and that their striving is kept within bounds by the need to use each other's research. The particular social dynamics in question are given in part by what Hull calls the "demic structure of science." What he says is quite interesting, but it is not that clear how he intends to pass from this description of the workings of science to the issue of rationality, which presumably involves normative considerations.

Hull cannot claim innocence by association with Lakatos. His use of the term "research program" has little in common with Lakatos', nor is there much connection with the context in which Lakatos spoke of rational considerations. For Lakatos, a research program is much more than the research carried out by one of Hull's "demes." Inter-deme disputes may still be intra-research program disputes. Moreover, the reason why priority disputes between research programs may be of rational interest is that leading in novel predictions constitutes a "progressive problem-shift." Lakatos' normative criteria placed emphasis on having theoretical and empirical momentum. Priority disputes of this sort would then help settle which of two competing "lineages" of theories (research programs) have more momentum. The view of the world articulated by the research program with the most momentum is most worthy of adherence and further development. The normative aspect enters precisely in this valuation. In

Hull's account, though, there is no question of norms guiding our choice of competing ideas about the world. There is one idea in common, the only question is who got it first. Hull's inter-group (or inter-deme) priority disputes simply try to settle which group won the race and should get the prize. Lakatos would thus find such disputes also to result merely from "vanity and greed for fame."

In spite of these objections, I think that in several respects Hull's view points in the right direction, and that his mechanism has some important things to tell us about how science works. I suppose that Hull speaks of rationality because he believes that the forces that lead to priority disputes are the same that account for the proper functioning of science. This point needs to be defended in connection with a larger view that permits us to say that attending to the proper functioning of science is the rational thing to do. Hull believes also that the peculiar objectivity of science results from competing groups of individuals keeping each other honest (by testing each other's research particularly when that research disagrees, and thus calls into question, their own research). As Hull puts it, "The greatest strength of science as it is now organized is that it harnesses our 'baser' motivations for more 'lofty' goals" (p.134). This approach suggests that properties like objectivity and rationality result from, or are identical with, organizational or social properties of science. Rationality would thus be removed from the hands of individual scientists. I may be reading too much into Hull's words, perhaps because I find this line of thought very congenial. In any event, it seems to me that this is the most fruitful direction to follow.

What we need, then, is a general view that permits us to tackle epistemological problems on the basis of biological ideas applied to science and that yields a social conception of scientific rationality. That is precisely the sort of view that I have been advocating for many years (Munévar, 1981, and recent papers, e.g., 1987 and Chs. 9 and 10). I cannot summarize that work here, but I will offer a very brief sketch to indicate the proper domain of Hull's selection mechanism and evaluate its worth to our understanding of science.

It seems to me that human scientific behavior, as all other human behavior, is part of the human phenotype. I imagine that Hull should agree since he says that "our propensity to learn about the world is surely based on our genetic make-up" (p.124), and that we are a curious species (p.124). He also makes several remarks about our genetic tendency to form groups and its contribution to the social character of science (p.124). Once we see this point clearly, however, we should not be as highly motivated, as Hull and so many other evolutionary epistemologists are, to account for the development of science (phenotypic and thus ontogenic) as if it were the equivalent of phylogenetic evolution (whether literally or by analogy). In Hull's case there is the additional complication of introducing a second "genotype" (or a "memotype" perhaps) where the standard genotype-phenotype distinction serves us well enough. (This problem he shares with Hahlweg and Hooker, 1987). I prefer my biological epistemology straight.

To that end, I have endeavored to determine the conditions that make science possible (not the logical conditions a là Kant, but the biological conditions). I have argued that science is a social expression of intelligence in dealing with the world and that intelligence depends on the brain, which of course results from a long natural history. Intelligence is characterized by its flexibility and its ability for indirect action. Intelligence also depends on the complexity of the central nervous system, for an increase in complexity is precisely what yields higher flexibility and more possibilities of indirect action. One aspect of intelligence, curiosity, is of particular relevance. Curiosity is a form of play, of play with the environment. At first sight, it is puzzling that curiosity should have evolved, since curious animals incur a great variety of risks for the satisfaction of a drive of no immediate pay-off. The reason why is that, in such free play with the environment, an animal develops many abilities singly or in combination which will later in life be of great advantage. In short, curious intelligence increases the thoroughness of an animal's interactions with an environment and its ability to deal with new or changing environments. This increase in adaptability can be illustrated by consideration of the many environments which humans occupy and from which they would be barred, had it not been for their curious intelligence. At a certain point the sophistication of our play with the world becomes social of necessity (just as we learned to hunt in groups we learned to satisfy our curiosity in groups). And here we have the beginnings of science.

From this account of the genesis of science we learn that by its very nature science has certain advantages to offer (as well as certain risks). To do science well is to increase our chances of receiving those advantages, and it is thus rational, in an old-fashioned sense of rational, to try to do science well (i.e. to attend to the proper functioning of science). From the same biological account we also learn that science is social by nature; and thus when we ask whether science is rational we should ask whether science as a whole is rational, not whether individual scientists are. I am making the same simple point that I would make in arguing that whether a team is good or not should be determined by observing the relations between the players on the field, not the individual quality of the team members outside of the context of the team play. The same point should apply to science since it is a social enterprise with divisions of labor.

Once we determine that scientific rationality is a social or organizational property, the same biological account will help us determine also how that property might be exhibited. For what we need to determine is whether science is so organized that it may increase its chances of realizing the advantages it has to offer. As we recall, one of those advantages is the ability to deal with new or changing environments. Now, scientific ideas are often developed to make sense of a certain range of experience. But the ideas that may do the job well in one context may not be appropriate to deal with a different range of experience. It pays, then, to have a strategy by which different approaches to nature are not only generated but developed. Ideally in my view, which I will not explain in detail here (see Ch. 9), science would show a variety of individuals or groups articulating sometimes different views of nature, sometimes different versions of the same view, at times with disregard of the views and even of the evidence presented by their opponents or competitors, but occasionally in contest with them. The sorts of considerations I adduce here are quite consistent, I think, with the considerations Hull adduces when speaking of the proper functioning of science and of the peculiar objectivity of science. But now those considerations can be interpreted within the framework of a more general biological view of scientific knowledge.

For epistemological reasons, then, science ought to exhibit this mode of organization. Not that individual irrationality becomes social rationality, but rather that scientific rationality, being an organizational property simply does not apply at the individual level. If science were to have the appropriate social or organizational structure, it should be considered rational, not only in accordance with the old-fashioned means-ends analysis of rationality to which I have already alluded, but because it would meet the conditions necessary to solve the contemporary problem of rationality in philosophy of science (Munévar 1984, 1987 and Ch. 9). A good understanding of the genesis and development of the conditions that make science possible yields a better understanding of the nature of science, and this in turn permits a more appropriate framing of questions about what science ought to be like. With these results in hand we have now a model against which we can compare actual science to see whether it is in fact rational. The answer to this question is a qualified "yes," I believe (Munévar 1984).

I do not mean to say that individual scientists are never rational; in many respects they have to exercise their good scientific judgments to choose the appropriate means for the ends that their daily work places before them. But what makes science rational is something else altogether. Nevertheless the behavior of individual scientists and of the groups they form is very important in other respects. My account of science is of a science of human beings, and therefore to be possible it must be consistent with the psychological and social nature of human beings. In other words, the naturalism I espouse cannot be complete until it includes an account of the dynamics of science at the individual and "deme" level. It would be a big mistake, however, to suppose that the role of the psychology and sociology of science is to describe the workings of science so as to generalize about how science ought to be. Different social conditions might make hitherto appropriate interactions between scientists or groups of scientists no longer helpful to the practice of science (Hull's half-hearted counterexample from the history of French science is not very convincing, for we might argue that those French scientists were too much men of their time and culture and it would have been difficult for them to adapt to a system so contrary to that time and that culture).

The role of the psychology and sociology of science should be instead to inform us of the manner in which different modes of psychological and social interaction contribute or detract from the organizational structure that science ought to have. And in this point Hull's mechanism does an excellent job of explaining how under a rather wide range of social conditions likely to appear in our civilization the very psychological and social make-up of human beings bodes well for the practice of science. For Hull's mechanism does not make scientists into any more than the flesh-and-blood human beings that they are, while telling us that, precisely because human beings are as they are, conceptual inclusive fitness and the demic structure of science will work to keep science honest. I do not buy everything that Hull has to say on this point, but I think that his valuable contribution is more than a step in the right direction.

I would like to close with two small points. The first is that the adaptability of science of which I have spoken ought to be ascribed not to individual scientific ideas, but rather to our capacity for science itself. Most scientific ideas have of themselves little practical utility, but they play a part in an enterprise which as a whole permits us to deal with the world in a manner that *can* have great utility (Munévar 1981, Ch. 4). The second point is this. I have rejected some of Hull's proposals, his analysis of selection, for example. And I have placed some of his proposals within a context more amenable to my own view of things. To manage this last I have given a sketch of my own biological epistemology, without elaboration or defense. In that sketch I have made reference to the crucial role of curiosity, a role that I have tried to explain. By not taking curiosity for granted, as Hull does, I think that some important aspects of the nature of science come to light. Now, there are two human drives that Hull takes for granted: the scientists' curiosity and their desire for credit for their contributions to science (p.154). In my own work I have placed great emphasis on explaining the first of those characteristics of scientists. As for the second, I would like to suggest a very tentative hypothesis. In my view (Munévar 1981, Ch. 4, and in a forthcoming book, *The Dimming of Starlight*, Ch. 3), I give an account of science as play; and I suspect that if we consider science in that light, it may be easier to understand why desire for credit, so prevalent in other forms of play, could achieve a significant role in the practice of science. Of course, desire for credit is present in many human activities, but Hull needs it to be in science more than in most others.

References

Hull, D. L.: 1988, "A Mechanism and Its Metaphysics: An Evolutionary Account of the Social and Conceptual Development of Science," *Biology & Philosophy* Vol. 3, No. 2.

Hahlweg, K., and Hooker, C.A.: 1988, "Evolutionary Epistemology and Philosophy of Science," *Issues in Evolutionary Epistemology*, SUNY Press.

Munévar, G.: 1981, *Radical Knowledge: A Philosophical Inquiry into the Nature and Limits of Science*, Hackett.

Munévar, G.: 1987, *Consensus and Evolution,* PSA 1986 2.

9 Science as Part of Nature

Understanding science as part of nature leads to a genuine epistemology. There is a widespread feeling among philosophers that any attempt to understand knowledge in naturalistic terms can be at best descriptive.[1] Genuine epistemology, however, tells us how we ought to go about the business of getting knowledge, and it is thus a normative or prescriptive discipline. But as we will see in this paper, that general feeling is quite off the mark. Coming to understand science as the result of natural history leads to the solution of important epistemological problems. The problem I have chosen as an illustration of the power of this naturalistic approach is perhaps the crucial problem of contemporary epistemology of science: the problem of the rationality of science.

That problem is this: Science was supposed to proceed by adherence to methodological rules that specified the means by which experience passed judgment on theory (e.g., providing inductive support for a particular theory, or falsifying it, and so on). But through the work of Kuhn, Feyerabend, and others, it has become apparent that no such rules are to be found, at least not general methodological rules. We should not take this to mean that no rationale can ever be given for preferring one scientific view to another, but rather that such rationale cannot be given by a set of rules of general application. If there is no statute law in science, to put the point in Lakatos's terms, the rationale will vary from case to case. If that is so, it seems that only those who are experts in the discipline will be in a position to see whether the correct decision has been made. According to Lakatos, this result would leave the matter of scientific rationality in the hands of an elite. And the trouble with an elite is that its procedures may degenerate for a variety of reasons (overconfidence, for example). The problem of rationality is, then, that we wish to have a scientific practice that manages to come up with rules, standards, or methods appropriate to each case, without having to depend entirely on the judgment of a possibly stagnant elite.[2] To the solution of this problem I now turn.

Let me begin by offering some motivations for thinking that science is part of nature. The first is that science is a social expression of intelligence in dealing with the world. And the second is that intelligence is a product of natural selection. Given these two considerations, it seems worthwhile to explore the possibility that science itself may best be understood within the context of natural history. This hypothesis, which surely needs clarification and argument, should be seen as belonging to the field of evolutionary epistemology. But to avoid serious confusion at the outset, it is advisable to distinguish my approach from others that my audience may have firmly entrenched in mind.

The evolutionary epistemology that Popper, Toulmin and others have made popular is basically an analogical view.[3] The epistemologist tries to show that the development of science closely parallels the evolution of species according to neo-Darwinism. Populations of scientific ideas, for example, shift under the pressures placed upon them by the intellectual environment. The basic problem with this approach is that after much impressive scholarship the epistemological job still appears undone. Let us suppose that the epistemologist succeeds in forging a close analogy. What follows from that? Surely not that science is rational. Being like the evolution of life, which is not itself rational, cannot suffice. A second problem with the analogical approach is that it cheats, a good deal of the time at any rate. The epistemologist wants to account for the history of science as a rational process. To help him in his endeavor he looks at the great success of neo-Darwinism in explaining natural history. He thus uses the prestige of neo-Darwinism in getting a hearing. But once he has made a few plausible comparisons, he rushes to announce that biological evolution is the responsibility of only one part of a much more comprehensive theory of evolution — a theory that includes such niceties as the inheritance of acquired characteristics, which cultural evolution is presumed to exhibit, or the coupling of variation and selection. In other words, the epistemologist brings back the very notions that neo-Darwinism had to discard in order to earn the prestige it enjoys today. Under close examination, thus, the analogical approach offers little in the way of justification or motivation. This is not to say that analogical reasoning has no place in trying to understand science; indeed, I will suggest later some ways in which it can be very useful. Nor is it to say that there is little to be learned from Popper, Toulmin, Campbell and those others who have taken up this approach. I, for one, owe them a great debt.

The approach I favor instead is straightforwardly biological. I do not wish to say that science is like nature, but that it is part of nature. To show this I should first provide a sketch of my account of the genesis of science. I claimed earlier that science was a social expression of intelligence in dealing with the world, and now I will try to make good on that claim. A characteristic aspect of intelligence, Piaget tells us, is that it allows organisms to transcend the immediate demands of their environment so that they may behave to their greater advantage at a more convenient time or place.[4] This indirect action of intelligence permits us, for example, to evaluate alternative courses of action on the basis of prior experience and to rehearse future actions in the imagination. Piaget found intelligence to be a powerful instrument of adaptation. This insight of Piaget's is buttressed by an analysis of the neural basis of intelligence. As the complexity of the central nervous system increases, so does the flexibility of its response. Information from the senses can now be rerouted, delayed, and stored; it can be compared with information from other sense modalities, as well as with previous information and with expectation. As complexity of the central nervous system increases, so does the number of modes of indirect action. Intelligence, of course, has many facets, but there is one fact in particular that makes clear how adaptability may be increased. I am referring to curiosity.

Jacob Bronowski once said that it is curiosity that liberates us from plain animalhood.[5] But as Konrad Lorenz pointed out, curiosity exists in many animals, in rats and ravens as well as in rational humans.[6] Curiosity is best seen as a form of play (with the environment), and as such it arises in situations that do not demand immediate attention to the environment. Animals play, or exhibit curiosity, not to satisfy hunger or sex drives directly, but because play (and curiosity as a form of it) provides a motivation of its own: It is enjoyable. In trying to satisfy its curiosity, an animal rehearses a wide range of skills, and of combinations of skills, that will later enable it to deal more effectively with the environment. For those skills, cognitive skills in this case, will permit the animal to either know its environment better or devise strategies by which to accomplish that goal (both in the sense of successful response; there need not be even a nod in the direction of realism here). Through curiosity, others can adapt to environments for which their species have not been "designed," and still others, who preserve much of their playful character throughout their lives, can adapt to changing environments. In the American Midwest, where I live, the cold temperatures alone would make it impossible for humans to survive year round, were it not for the fact that humans have applied their intelligence to produce clothing and shelter, to invent fire, and to protect themselves from the inclement weather in many other ways. To do even better than that, humans have applied their science and technology and are thus able not only to live rather comfortably but to seek occasional respite from their environment in the benign climates of lands as far as Australia.

The evolutionary justification of curiosity, and of play in general, can be found in the increased range of abilities that become useful later in life. But there is, of course, a cost to the individual. When playing with others or with the physical environment, an organism is not eating, it is not mating, and it is not doing anything of immediate use. To make matters worse, the organism is distracted by its play and may fail to notice the approach of a predator, or some other danger. The payoff is often long in the future. This postponement of material satisfaction is a wonderful illustration of the indirect action that intelligence permits. In any event, my suggestion is that we can find the origins of science at the juncture where human curiosity about the world becomes social.

It seems to me that, at least for humans, the relentless attempt to satisfy our curiosity about the world inevitably becomes social. Just as we came to hunt in groups — an exercise in another form of social intelligence — now we try to satisfy our curiosity in groups. There are two main reasons why this should be so. The first is simply that to explore our environment in great depth often requires the cooperation of others. An experiment in gravitational physics may have to be carried out beyond the Earth's atmosphere and will involve fields as diverse as rocketry, metallurgy, superconductors, chemistry, and orbital dynamics, as well as the general theory of relativity. Even in the same field some projects are much too large to be taken up by a single investigator. At a certain level of sophistication, division of labor becomes of the essence. A second reason is that the very attempt to satisfy one's curiosity may require the prior existence of an institution devoted to such a goal. One cannot just decide to study the most basic components of nature unless one has entry into a society committed to a program of research in elementary particle physics — likewise, one cannot just decide to become a milkman in a continent where placental mammals do not exist.

Once it becomes social, the attempt to satisfy our curiosity about the world gains extraordinary power, and so do the skills that result from it. Now, if this general account is correct, we should expect that such a social enterprise would allow us to *(a)* have a more thorough interaction with the environment, *(b)* increase the number of environments with which we can deal, and *(c)* deal with a changing environment. This social enterprise is, of course, science. And I have already suggested how science indeed provides such advantages (e.g., contemporary living in the Midwest). I do not mean to suggest that in every case when a human skill takes on a social character it thereby becomes more effective. It was so in the case of hunting in the environments our ancestors faced until a few thousand years ago. And it seems to me that it is so also in the case of science because a compelling biological analysis of intelligence lends itself naturally to this social extension.

What science has to offer, then, is a more thorough interaction with the environment and an increasing number of environments, including changing environments, with which we can deal. This is the function that science may perform for us, if we choose to engage in it. Success in the second of these achievements does not guarantee success in the first. Because of its ability to deal with a greater range of environments, a species may move into a particularly rich environment. But subsequent changes in the environment may be so drastic that this particular species becomes extinct, in spite of its flexibility. Groups and individuals undergo similar tribulations. Because of curiosity many animals will later accrue great advantages. But at an early stage some of them will die because of perils that must be blamed on their curious nature. That is all I will say here about concerns with nuclear weapons and other monstrous creatures of modern science; I have explored the matter in greater detail elsewhere.[7]

It would be a mistake to think that organisms incapable of developing science cannot also be very adaptable. What increases adaptation (or the potential for adaptation) in one type of organism depends on the sorts of interactions organisms of that type can have with the environment. Thus an opposable thumb may be of great value to animals that exhibit a certain skeletal structure and development of their central nervous system, e.g., humanoids, in many but not all environments. But to other types of animals, say cows or butterflies, an opposable thumb would be disadvantageous in most typical environments. Likewise with intelligence: The increase in metabolism that goes hand in hand with an increase in the complexity of the central nervous system carries a price tag that may be too high for many organisms. A move in the direction of high intelligence, let alone its social telescoping into science, may not even begin.

Nor should one think that every product of science, or every scientific skill or technique must be clearly adaptive. Surely the model of science as arising out of curiosity does not entail such a conclusion. After all, not every skill that an animal develops in his play with the environment will later prove to be of the greatest usefulness. Some of them will be of no use at all. And others will be put to use indirectly. After all, if science is play, as I suggest, we are likely to devise all sorts of games in exploring our world. And some of those games are bound to be very abstract and intricate. But a few of those may greatly facilitate the application of some other skills in our dealing with one or several environments (e.g., by providing for conceptual, mathematical, or instrumental elaborations of our theories).[8] Kuhn already pointed out that much scientific work goes into the articulation of the main views we hold.

With this account of the genesis and nature of science in mind, let me turn now to the problem of rationality. Given that science is a communal enterprise with division of labor, the question of the rationality of science should be asked of science as a whole. This point goes directly against the typical manner in which philosophers have approached the question of rationality: they look at whether this or that great individual scientist, or research group, adhered to this or that set of methodological rules. But it seems to me that to approach the question as philosophers have done is to commit a logical mistake. In trying to determine whether a football team is good, we cannot merely look at whether its players are individually good. We wish to know the social or structural relations that the team exhibits during its games, whether, in short, it can play as a team. When a player creates space into which another can move to receive the ball and score, the social unit is working well. Even brilliant individual action often depends on good positioning by teammates that keep the defense guessing about what the next play is going to be. In any event, to ascribe the properties of the individual members to the whole team would be a mistake. And it seems to me that the same is true in science.

I propose that the question, "What would it take for science to be rational?" should be thought to be equivalent to the question, "How should science be structured so as to perform its function?" My evolutionary account forces us to pose the question in this way, and it also suggests how to answer it. To determine how science should be structured or organized so as to perform its function is to determine what it would take for science to enable us to adapt to new environments or to a changing environment, and so on.

We may then easily realize that scientific views are often designed to make sense of a particular environment: that of our experience. But success in one environment, or in one context, does not guarantee success in others. If the environment or context is likely to change, it pays to have a strategy for generating alternative points of view. That is, an organizational requirement of science is that it allows dissension and the generation of alternatives. This requirement of intellectual freedom must be accompanied by another. Scientific views must be given a chance to develop. They have to begin like all ideas: small, and almost certainly vague. But if we see some promise in them, we should not abandon them just because they are in conflict with the accepted ideas of the time, and not even because they are in conflict with the evidence. We may do so, but we should not have to. Otherwise, ideas would never blossom into glorious scientific achievements. (The reason counter-evidence need not always be decisive is that observation and experiment always have to be interpreted, but the interpretation that makes them into counter-evidence may depend on theories that the very development of the new views would expose as inadequate).

These two requirements of intellectual freedom, that science must be so organized as to permit and perhaps encourage the generation and development of new ideas, must be met by science as a whole, not by individual scientists. Some scientists will be generating new approaches, others will be developing them in a very stubborn fashion, and still others will reject any but the accepted views of the time. Some scientists will be open-minded and some will not. It does not matter as long as there is enough room in science for all kinds. If there is, if science does employ a strategy for generating and developing new ideas, then science will be in a better position to adapt flexibly to new challenges. It will thus permit us to deal with new or with changing environments. If so, it will perform its function, the function suggested by my biological account. And in a very straightforward means-ends analysis of rationality, we should conclude that science would then be a rational enterprise.

Notice that this means-ends analysis also provides a recipe for solving the contemporary problem of scientific rationality. Two demands were placed upon the epistemologist. The first was that science proceed in such a manner that its practitioners generate opportune methods and procedures. When viewed from the perspective of my social conception of rationality, science offers precisely a general strategy to improve the chances of accomplishing the desired goal. The second demand is that the first should be met without tying science to the dangers of being ruled by a stagnant elite. The two requirements of freedom under the social conception will reduce such dangers. After determining what science ought to be like, however, we still wish to know whether science is actually rational. I think that it is largely so. Even under the most challenging description of the history of science, i.e., Feyerabend's, we can see that science exhibits the required strategies. Indeed, what I have called the two requirements of intellectual freedom overlap to a great extent with Feyerabend's principles of proliferation and tenacity.[9] What looks like anarchy under a conception that equates rationality with adherence to methodological standards, now looks like the very sort of organizational structure that science ought to have. With the shift to a social conception we also shift from looking for rationality in the choice of theory to finding rationality in the ability to reach certain goals. As it happens in science itself, the solution of a problem takes place within a transformation of outlook in the field.

I do not mean to say, incidentally, that irrationality at the level of the individual becomes rationality at the social level. My point is rather that the concept of scientific rationality no longer should be applied to individual scientists. Social properties are social properties. Nevertheless, there are many other ways in which the question of individual rationality may still come up. For example, once a certain view of the world is found promising, by a scientist or group of scientists, procedures are devised for developing it further and for testing it. Many goals and subgoals must then be reached, and some means may be more effective in reaching those goals. Once more we would use a means-ends analysis of rationality.

In the transformation of epistemology that ensues from taking seriously the idea that science is part of nature, not everything is entirely new. Many readers will no doubt have already noticed the similarity between my requirements of intellectual freedom and the conditions for the growth of science put forth by other philosophers, Toulmin for example. The crucial difference is that whereas for Toulmin such conditions lead to the evolution of populations of ideas, for me they lead to an increase in the adaptability of the human societies that practice science. This shift to a truly naturalistic epistemology can better place the work of many other epistemologists with social and naturalistic inclinations. For we all wish primarily to understand science, not to draw philosophical distinctions that protect the autonomy of epistemology at the price of keeping the nature of science beyond our grasp.

Once such a distinction, between description and prescription, seemed to doom naturalistic approaches to something less than a genuine epistemology. But now we can brush this qualm aside and get on with our naturalistic programs. One task that still must be done in much greater detail is the determination of the social, organizational, or institutional dynamics by which science can best perform its function. For example, the interplay between the two requirements of intellectual freedom, striking out for novelty and sticking to one's point of view, is not restricted to big groups of scientists at odds with other groups. Such interplay takes place in many ways, and some of them are described, for example, in several papers in *Issues in Evolutionary Epistemology* (see Acknowledgements). Hooker, for example, tells us about hierarchical levels, or layers, with a variety of feedback loops. In his model, individual scientists will fill one or several institutional roles, but in doing so they will find themselves sometimes in cooperation and sometimes at odds with other scientists with different institutional roles. For instance, a team of physicists working on the same project may well represent different areas of expertise within physics. In his institutional role, or in his role within the group, the plasma physicist should push his particular perspective as to what is important, and so should the particle physicist. The mathematical physicist will have other concerns that the group may face sooner or later. Hooker is right in suggesting that we will find some overlap as well as some different ideas on what to do and how to interpret what has been done. When the research group interacts with another research group, as Küppers tells us, I imagine that we will find many other opportunities for variation and recombination, as well as for preserving the view initially favored by the group. I purposely chose a biological analogy here, for I think that in tasks such as these analogical reasonings has its greatest role to play. I have in mind, for example, the work of David Hull. In particular, I find very useful his accounts of he way in which rewards in science help achieve cooperation and keep the discipline honest.[10] We must bear in mind, however, that even though analogies suggest hypotheses about the mechanisms that determine the social dynamics of science, those hypotheses will eventually have to stand on their own.

Traditional philosophy is not very likely to take kindly to my approach. I cannot attempt to refute every conceivable objection in the space of one chapter, but let me go over one of the main sources of discontent: the feeling that I am treating rationality as the product of an invisible hand mechanism and that there is something terribly wrong about it. One way to express this feeling is that on my account scientific rationality would not be the product of conscious deliberation (since science as a whole seldom deliberates), and thus it is inappropriate to speak of rationality. If a means-end analysis is to be employed, we had better know what the ends are supposed to be and how the means will get them for us. Nevertheless, it seems to me instead that this very demand for conscious deliberation is inappropriate. Philosophers of science have been trying to determine for centuries what scientists ought to do, or what they do when behaving close to ideally. But these philosophers have been seldom deterred by what scientists announced their method is, for often scientists say one thing but do another — and they may do the wrong thing anyway. The "real" scientific method was then not part of the public domain, surely not a matter of conscious deliberation. As Lakatos put it, the philosopher's job is to discover the universal criteria that great scientists applied sub- or semi-consciously in their evaluation of particular cases.[11] I do not believe there are any such universal criteria, but the point is that mainstream philosophy of science has always found it comfortable to speak of rationality in the absence of conscious deliberation. If this is an objection to my view, then it is an objection against doing mainstream philosophy of science at all.[12]

A second way of expressing the nagging feeling is that if science is part of nature then it is somehow inevitable; and if it is inevitable, what is the point of talking about rationality? It seems to me, however, that our natural talents may or may not be developed: We often choose what to do about them. As basic a skill as eating may or may not be properly exercised (just look at practically any crowd of human beings in any Western city); or we may choose not to exercise it at all and starve ourselves to death out of spite or political passion. That our ability to produce science is a product of natural history need not prevent our choosing whether to practice it more, less, or not at all. We may be lazy or not daring enough. Moreover, science would be one of many social skills. In many social situations, science would not arise because it would conflict with and bring into question the society's mechanisms for preserving the cohesion of the group (e.g., religion). And there may be other conflicts as well. Once we see science as part of nature, however, we can understand its nature better, and we may also frame better questions about its rationality and its worth.

This brings us to a very important point. I speak of science as rational because the ends it permits us to achieve are presumably worth achieving. I have spoken of how science permits us to deal with new or changing environments, of adaptability in short. Given my naturalistic account, it seems that science has some worthwhile function to perform for us. But is that function truly worthwhile? Should adaptability take precedence over other goals that human beings may also have? The question of the rationality of science thus leads very naturally into questions about values, not just epistemic values but also social and moral values. In this manner, my naturalistic account brings epistemology and ethics together. Some hard-nosed epistemologists would find this result a matter for despair. It is bad enough to worry about the rationality of science without also having to fight our way out of the morass of value. But I think that the situation is much more encouraging than that. With the standard inferential models with which philosophers have tackled questions of value, we could not get very far. But I have proposed elsewhere a causal theory of value in which knowledge and value form an intricate network that can be criticized in a variety of ways.[13] Although a fuller account of this theory will be given in Part III, let me point out the moral of this story: A naturalistic account thrives not by ignoring the normative aspect of human thought, but by joining that aspect with the rest of our nature.

I trust that what has been said so far establishes evolutionary epistemology as a genuine epistemology. I would like to mention briefly other areas of epistemology where a truly evolutionary approach has important contributions to make. One of them is the issue of realism, in which evolutionary considerations can be powerful philosophical tools. As I see it, a truly evolutionary epistemology commits us to a very sophisticated relativism, but I will concede that some very sophisticated versions of realism may also be inspired by it.[14]

A second area is in our understanding of the origins and motivations of science. It is commonplace for philosophers to say that science has its origin in problem solving. Popper and others have stressed this point throughout their work.[15] But this commonplace can be misleading. As I have argued, and as scientific intuition often demands, science has its origin instead in curiosity, in our play with the world. Science is essentially play, a game. Within games there are always problems to solve, but it seems to me that to speak of problem-solving, in this sense, is not very informative.

A third area directly affects some of the concerns expressed by contemporary evolutionary epistemologists. For at least two decades now, epistemologists have been trying to explain the history of science as rational, with mixed results, to be charitable. To some, evolutionary theory appears full of promise in tackling this task. But if we take evolution seriously enough to see science as part of nature, the epistemological task may begin to look like an unnecessary compulsion. For science should then be seen as part of the human phenotype, as other behavior surely is. Consider organisms as simple as bacteria. In a poor environment, one type of bacteria will prey on their competitors. If the environment becomes rich in nutrients, the bacteria will change their behavior radically: they will move away instead. We can explain their behavior by reference to their internal organization, their past history, and so on. But this explanation will go outside of the description of the two phenotypic states, of the history of the behavior itself. The second behavior does not "follow" from the first in a logical or rational way. The organism simply undergoes a radical change of posture toward the environment. What makes sense of the change is the knowledge of the nature of the organism and the character of its interaction with the environment. And it appears to me that we cannot exclude the same kind of consideration in the case of human science.

If science is part of the human phenotype, there need not be logical or rational continuity from one stage of its history to the next. Satisfying our curiosity about the world may demand, on occasion, a radical change in our approach to the game. The behavior of the small infant need not be continuous with that of the older child, although the change from one to the other may be explained by Piagetean considerations about ontogeny. In a similar fashion, the content of our science at a certain stage in development need not be continuous with that which follows it (although in many actual cases there could be quite a bit of continuity). For under radical pressures or the excitement of truly radical new ideas, science may switch just as radically from one point of view to another. The rationality of science should not depend, then, on the logical continuity of the content. We may find it instead outside of that history: in the mechanisms that bring about science and in the function that they allow it to perform.

Notes

1. M. Bradie, "Assessing Evolutionary Epistemology," *Biology and Philosophy* 1 no. 4 (1986): 401 — 59.
2. For more details on this problem see Chapter 7.
3. K. Popper, *Objective Knowledge*, (Oxford: Oxford Univ. Press, 1972); S. Toulmin, *Human Understanding*, vol. 1, (Princeton: Princeton Univ. Press, 1972).
4. J. Piaget, The Psychology of Intelligence, (Totowa, N.J.: Littlefield, Adams & Co., 1972).
5. In his famous television series, "The Ascent of Man."
6. K. Lorenz, *Studies in Animal Behavior*, (Cambridge: Harvard Univ. Press, 1971). See also my *Radical Knowledge*, Ch. 5, (Hackett, 1981).
7. *Ibid.* This point is worked out in greater detail in my upcoming *The Dimming of Starlight*.
8. For details see *Radical Knowledge*, Ch. 4. The reference to Kuhn is to his famous *The Structure of Scientific Revolutions*, (Chicago: Univ. of Chicago Press, 1970).
9. P. Feyerabend, *Against Method*, NLB (1975), and "Consolations for the Specialist", in I. Lakatos and A. Musgrave, *Criticism and the Growth of Knowledge*, (Cambridge: Cambridge Univ. Press: 1970).
10. D. Hull, "Altruism in Science: A Sociobiological Model of Co-operative Behavior Among Scientists," *Animal Behavior* (1978):26.
11. I. Lakatos, "The Problem of Appraising Scientific Theories," in his *Mathematics, Science and Epistemology*, (Cambridge: Cambridge Univ. Press, 1978).
12. An independent argument can be found in Chapter 7.
13. See my review of P. Singer's The Expanding Circle in *Explorations in Knowledge*, vol. 4, no. 1 (1987): 43 - 50.
14. See my *Radical Knowledge*, Chs. 2 and 3; and C. Hooker's *A Realistic Theory of Science*, (Albany, N.Y.: SUNY Press, 1987). See also Part I.
15. K. Popper, *Objective Knowledge, op. cit.* 242-44.

PART III
FROM EPISTEMOLOGY TO ETHICS

10 Evolution and Justification

In tackling the issue of justification, philosophers have gone overboard in their efforts to discover fallacies in the reasoning of otherwise sensible people. And having fallen into deep water, those philosophers now find that they do not know how to swim. The cause of such philosophical distress can be located in what may have once appeared as a virtue: the separation between man and nature. Science may describe the world, and even us, but it cannot tell us what we ought to do; it may explain how a state of affairs originated and developed, but it cannot tell us whether that state of affairs is a good thing. In this paper I will argue against that separation; and in advocating a philosophy that treats man as part of nature, I will attempt to throw light on the issue of justification.

The typical philosophical move towards the separation begins with innocent plausibility. As a philosopher once said to me, "My mother taught me to judge people by their actions and not by their origins." This sounds so right that we draw a moral for the worth of our ideas about the world: the genetic fallacy. To that moral we add the dictum that to describe and to prescribe are distinct; and we can now hold, say, that the history and sociology of science are irrelevant to its justification. As Hempel put it, the philosopher should only be concerned with "certain logical and systematic aspects of science which form the basis of its soundness and rationality . . . in abstraction from, and indeed to the exclusion of, the psychological and historical facets of science as a social enterprise."[1] This distinction is just as prevalent in ethics, where it also places sharp limits on the philosopher's attempt to solve the problem of justification. Although the reach of the distinction extends even further, I will limit myself in this paper to two areas: the epistemology of science and the resolution of conflicts of values. As we shall see, these two areas are not independent of each other.

Let me begin with the epistemology of science. I will first indicate how the main problem in this area arose from the failure of empiricism, and then I will argue that a naturalistic account can solve that problem. It will then be clear that naturalism yields a genuine epistemology.

When the logical approach was challenged by historically minded thinkers, a typical response was that epistemology should be concerned with methodological norms. The philosopher's job was to determine what scientists ought to do; and thus considerations about what scientists actually do were not relevant, strictly speaking. This typical response, however, was bound to fail for several reasons. One of the most important originates in the requirement of empiricist epistemology is that experience ought to take precedence over theory. This requirement would seem pointless, of course, unless we assume a further distinction between theory and observation (the mode of experience by which the world presumably has a say in our choice of theories — this is what made "empirical" science empirical). As is well known, the examination of the history of science, by Kuhn, Feyerabend and others, created serious problems for this distinction.[2]

Let us consider just one example: the Tower Argument against the motion of the Earth. If we drop a stone from a tower, and the Earth moves, the tower will have moved by the time the stone hits the ground. There will then be an appreciable difference between the distances from tower to stone at the beginning and at the end of the experiment (remember that the Earth rotates 1000 miles per hour). But no such difference could be detected. For the stone to keep the same distance it would have to fall obliquely. It is plain to the eye, however, that the stone falls straight down. This fact about the motion of the stone defeats the idea that the Earth rotates. Nevertheless Galileo argued that there was no such fact: the stone does not fall straight down, it only seems to. The real motion of the stone is given by two components, the circular inertia of the Earth, which the stone shares with the Earth, the tower, and the observer, and a vertical motion towards the center of the Earth. The observer detects only the vertical component precisely because he shares the circular inertia of the Earth (just as when traveling in an airplane, we do not observe the person sitting next to us to be moving at 600 miles per hour, but we do observe the nodding of his head as he falls asleep). As we can see, even "facts" as plain as the vertical motion of a stone can be interpreted differently if we make different theoretical assumptions. A scientist can save his view by specifying a different "empirical basis." Therefore, experience does not *always* take precedence over theory.

The great problem for empiricism is not that scientists often fail to live up to the methodological norms discovered by philosophers, but that they have to — if science is to advance, that is. The reason is simple. In many of the most admired episodes in the history of science, the winning view was often in trouble with the facts and had to be rescued by challenging their verdict. To have adhered to method would have *prevented* the great accomplishment — not merely delayed it. From the scientist's point of view, then, epistemological purity is simply absurd. The epistemologist saw rationality as adherence to methodological norms. For the scientist, rationality was to be recommended because those norms would improve his prospects of coming to understand the world. But when it turns out that adherence to the norms may actually hamper those prospects, the scientist must make a choice between rationality and scientific success.

Since the rationale for rationality was that success, the scientist's choice is obvious.

Nevertheless many philosophers, particularly the logical empiricists, took up the curious notion that epistemology should not have to give advice to the practicing scientist. The job of epistemology was to "explain" science. This "explanation," which was attempted by means of rational reconstructions, was patently aprioristic; but in the last analysis such an approach was a pretence. On pain of absurdity, the logical philosopher had to put forth norms consistent with procedures that secure knowledge about the world. But how could he discover such norms? I submit that he "abstracted" them from what he understood of the practice of science. This Hempel has come to admit: Explication and the like "were never undertaken in a purely *a priori* manner . . . [they] were constructed with an eye on the practices and the needs of empirical science."[3] The philosopher makes up theories about science just as the scientist makes up theories about the world. In this the logical approach turns out to be indistinguishable from the historical one. Otherwise the main difference is that whereas the second treats the practice of science seriously, as given by socio-historical studies, the first scorns it publicly. It is not surprising that the historical approach was eventually seen as more promising in our wish to understand science.

The triumph of the historical approach already undermines the use philosophers make of the separation between description and prescription: matters of practice can be used to pass judgment on matters of norm. That triumph, however, does not take us far enough in questions concerning justification. We might think that all we need to do is base our norms on the actual successful practice of science and we will be able to justify our future decisions by reference to such norms. The problem with this suggestion is that, as Feyerabend and Lakatos point out, a practice may degenerate.[4] Moreover, norms based on practice require that we take the word of an expert elite; for only experts will be in a position to determine whether the changes in the practice are for the best. This situation brings up one of the many ways in which a practice may degenerate: Overwhelming success may drive the elite to complacency and over-confidence.

Since universal norms seem unlikely, given the breakdown of the distinction between theory and observation, we may conclude that different methods and procedures would be appropriate in different circumstances. As epistemologists, we must now determine whether science has a strategy that will increase its chances of coming up with such methods and procedures when circumstances demand. And all the while we must guard against simply trusting the judgment of a possibly stagnant elite. This is the problem that the new epistemology of science must solve.[5] In that solution lies the key to the issue of justification.

This is also the problem that a naturalistic approach can solve. Of course, not all naturalistic approaches will do. What I have in mind is a naturalism that begins with the genesis and development of the skills that make science possible. If this plan sounds like a recipe for committing the genetic fallacy, so much the better; for we will then come to realize the limitations of the philosophical injunction against considerations of origin. We will come to see that in understanding the genesis of science we will understand its nature; and that in understanding the nature of science we will find the means to settle the question of its rationality. Since the epistemology that follows is both naturalistic and evolutionary, I would like to point out briefly how it differs from views that the reader may tend to conflate with it: standard evolutionary epistemology and Quinean naturalized epistemologies.

Once philosophers learned well that universal standards or norms do not rule the manner in which the content and procedures of science change, their efforts were directed toward showing that the development of science was rational. To this end it occurred to Popper, Toulmin, and others that the history of science was *like* natural history.[6] Scientific ideas, or populations of scientific ideas, evolved somewhat in the manner in which neo-Darwinism says that populations of organisms evolved. This analogical approach fails to meet my purpose on several counts. First of all, it is not really naturalistic. Human thought and nature still go their separate ways. In Popper's account, for example, the selection of ideas takes place not in nature but in some metaphysical Third World that rivals Plato's Forms. Second, even if a close analogy holds, neo-Darwinsim does not entail that the history of science is rational. Darwin himself made it very clear that there was nothing rational about the evolution of species. I do not wish to say that science is *like* nature but that it is *part* of nature.

Apparently, however, a straightforward naturalism must face anew the problem of the distinction between describing and prescribing. For a naturalistic epistemology is presumably descriptive, and no matter how good the description is, a logical gulf would still separate it from the normative domain. Therefore a naturalistic epistemology would have little to tell us about justification. In this argument we find a source of disenchantment with Quine's proposal to naturalize epistemology.[7] Granted that he was right in suggesting that, since the positivists' logical reconstructions of scientific knowledge were dismal failures, it made more sense simply to inquire into how science is actually generated. But a philosopher may still want to ask whether that science is rational, no matter how many Wittgensteinian appeals he hears to grow up and forego certain questions.

I think that the urge to ask such questions is properly philosophical; but in opposition to the common wisdom, I also think that naturalism is the way to satisfy it. I start from two plausible conjectures: that science is a social product of intelligence, and that intelligence depends on the brain. Since the brain is to be explained by evolutionary biology and neuroscience, I shall take my cues from those two sciences. Intelligence arises out of perception and other biological structures and its adaptive features are directly connected with the characteristics that distinguish it from those structures. Intelligence, as Piaget has remarked, allows us indirect responses to the world, it frees us from having to respond merely to the immediate and pressing demands of the environment.[8] This psychological account of intelligence accords well with our understanding of the evolution of its neural basis. As the complexity of the central nervous system increases, so does the ability to treat information in a variety of ways. For such complexity often includes many connections between different sense modalities, as well as structures to coordinate the senses with other functions of the central nervous system. The result is that in humans, for example, we do not simply see with our eyes but with our whole bodies. What we see depends on internal senses that factor the position of the body in perception (so the image of our audience remains stationary even though we move), on what we smell, taste, or hear (as the shapeless bundle in a dark street instantly becomes a guard dog when we hear that nasty growl). What we see also depends on structures of the brain that aim at constancy. Thus, for example, we see a circle that is partially turned away from us still as a circle and not as an ellipse. Expectation and the success or failure of behavior also play a major part. For instance, a subject looking at a distorted room (in which the left wall is much longer than the right) through a peephole will see it as normal — square, that is. When he is asked to touch with a stick an object on the wall, he will fail repeatedly; but then suddenly his perception of the room will change and he will see it as it is. We realize both at the psychological and neural level that information can be rerouted, stored, compared and transformed in a great variety of ways.

This complexity of the central nervous system underlies the ability for indirect action. And nowhere is that ability better illustrated than in curiosity. Curiosity is a form of play, more specifically of play with the environment. In exercising its curiosity, as in other forms of play, an organism incurs a variety of risks. In general the curious animal is not responding to immediate pressures of the environment: It is not eating, mating, escaping, or looking for shelter; nor is it paying enough attention to the possible predators. Why should curiosity have evolved then? The most sensible answer, it seems to me, is that in spite of the initial risks, curiosity offers a great pay-off later in life. In their play with the environment, rats, ravens, apes, and hominids rehearse many skills, singly and in combination, that may later be of great use in dealing with their environment. In play for its own sake we enhance our bodies and central nervous system's ability to deal with a great variety of circumstances. And playing with the environment so as to satisfy its curiosity enables an organism not only to get to know its environment in ways that will prove useful later, but more importantly, to develop strategies by which it can come to know its environment better.

This flexibility of response pays some clear dividends. First, an animal, as we have seen, may come to know its environment better. Second, a species of animals can move into new niches, for its curious young can then adapt to a wide range of environments. Third, organisms which remain curious throughout their lives can better deal with a changing environment. Curiosity, in short, makes organisms more adaptable.[9]

I do not mean to suggest that curiosity is the only way by which an organism can become highly adaptable. A characteristic is adaptable or not in part because of the way it fits in with other characteristics. An opposable thumb is adaptable only for animals with the appropriate skeletal structures and neural coordination. It was a considerable advantage for our ancestors, it would be a hindrance to horses, and a fly would have no place to put it. Likewise, the increase in complexity of the central nervous system required for curiosity is adaptable only for animals that can afford the concomitant increase in metabolism. In some cases an increase in size is necessary as well. But that increase may be fatal to animals that survive by hiding from their predators in small places. An increase in the complexity of the central nervous system served our ancestors well because of the kinds of animals they already were. But for different kinds of animals natural history can provide other routes to adaptability.

As the sophistication of curiosity increases, the attempt to satisfy it becomes science eventually. In humans, intelligence takes many social forms, and curiosity is no exception. Just as we came to hunt in groups, we have also come to satisfy our curiosity about the world in groups. In this social extension of intelligence we find the genesis of science. That the attempt to satisfy curiosity becomes social is not surprising. Often the task is just too large or demanding for a single individual. Not only greater numbers are necessary, but a division of labor must be instituted (in contemporary science many projects require an ensemble of experts from many different fields). Furthermore, certain lines of investigation cannot even be started by an individual unless there is a prior institution of science committed to his training and employment in certain ways of viewing the world (could a member of an isolated hunter-gatherer society devise means of detecting heavy leptons? How could the thought even occur to him?)

This social telescoping of curiosity increases, it seems to me, the already considerable adaptability offered by intelligence. Most of the readers of this journal live in the Northern Hemisphere, at latitudes where beings identical to *homo sapiens* in all respects but intelligence could not possibly survive. Tall, hairless bipeds like humans needed intelligence to make fire, clothing, and shelter. Intelligence permitted them to move into the new environments of the cold north and to adapt when those environments changed as the ice ages came and went. Today social intelligence, in the form of science, permits their descendants a far greater degree of comfort and many new ways of interacting with the environment. It also permits them to move into hitherto forbidden environments, such as Antarctica and outer space, and in general to expand the niche of humankind.

The social satisfaction of curiosity, like the individual one, brings about also an increase in risks. But the new risks do not preclude adaptability in the social case anymore than in the individual case. Many a young animal has perished in its pursuit of curiosity, but we may still say that curiosity made that *type of* animal more adaptable. On the whole the advantages are there. A full argument for the advantages of science would be beyond the scope of this paper (although I have taken it up elsewhere).[10] In any event, as we will see, such an argument is not truly necessary for the purposes of this paper. Let me restrict myself, then, to the biological notion of adaptability: a more thorough interaction with the environment, and the ability to deal with new or changing environments. These are *prima facie* advantages — and science does offer them.

This evolutionary account of science does not imply that science is somehow inevitable. Many of our natural talents go undeveloped, in spite of the advantages they might confer upon us. Science need not be an exception. We could do more science, we could do less, perhaps we could do none at all. We could do science badly, we could do it well. Since doing science well would confer the advantages already mentioned, it seems only sensible that we should wish to do it well. Indeed, it would be rational to do so, in a straightforward means-end analysis of rationality.

To do science well would mean, in my account, to do it so that science will permit us to deal with new and changing environments, and so on. To determine how science can do this is to determine what it would take for science to be rational (since it would specify the means by which science can attain the desirable ends it has to offer). The first step, given my evolutionary account, is to treat seriously the notion that science is essentially a social enterprise. Accordingly, the question of its rationality should be asked of the scientific enterprise as a whole. To try to settle this question in the traditional manner, i.e., by examining the behavior of individual scientists, is a logical mistake. The rationality of science should not depend on whether this or that famous scientist adhered to this or that set of methodological norms. We should rather be concerned with whether science is socially constituted so as to achieve its ends. Similarly when we wish to determine whether a basketball team is good we look at the organizational and social relations that the team's members exhibit during the game. As any sports fan knows, the team with the most talented individuals on its roster need not be the best.

Just as my evolutionary account indicates that rationality ought to be a social property of science, if it is a property of science at all, it also suggests what the social or organizational structure of science ought to be. Scientific views are characteristically built up with the aim of explaining some area of experience, and thus they are constrained in many important respects by our range of experience (even though they always have consequences beyond that range). In my jargon, scientific views are designed so as to fit a certain environment (or rather to allow us to fit a certain environment). But ideas that work well in one domain of experience may be completely inappropriate in others. If we are to deal with new and changing environments, it pays to have at our disposal a variety of ideas and modes of thought. In this manner science can better meet the challenge of attaining its ends. Science must then be organized so as to permit the generation of alternative ideas.

Nevertheless the freedom to generate new ideas, and perhaps the encouragement to generate them, is not enough. Ideas are born small; they need time and effort to develop into full-fledged alternative points of view. Some scientists must then be permitted the freedom to develop those ideas until they can compete with older views. That is, science must permit that some of its members stick to their views in the face of intellectual unpopularity and even in the face of contrary evidence. The reason is twofold. First, as we have seen, a view in trouble with the facts may still be rescued by the generation of a new empirical basis. Second, even when the scientist is not able to generate a new empirical basis outright, he might take comfort from the following thought: The observations used to judge his theory are interpreted in some auxiliary science (e.g., hypotheses about astronomy are judged on the basis of complicated telescopic operations that span a variety of fields in physics, chemistry and computer science — the distinction between main and auxiliary science is like that between foreground and background: it shifts as one looks elsewhere). These auxiliary sciences have reached an accommodation with the main view, normally after long struggles. A truly radical new view is thus likely to conflict with views in equilibrium with its established rival. It seems, then, that disagreement with the facts can be a sign of progress. This second requirement of scientific freedom stands Goodman's plea for entrenched theories on its head; but it should not be taken to mean that all views are on a par. The scientist who is permitted to work is still ahead of him. Most hunches will lead nowhere; and it is not easy to develop a hunch to the point that the original vision becomes commanding enough for others to share. Hard work and ingenuity in a climate of freedom, however, will provide science with the flexibility it requires to allow us to deal with new and changing environments. That is to say, if science is organized so as to meet these two requirements of freedom, on the generation and development of alternatives, it will increase the chances of attaining its ends.

My evolutionary account thus solves the contemporary problem of the rationality of science. For in the optimal social organization of science we find a strategy to increase our chances of developing appropriate ideas, procedures and methods to meet the demands of changing circumstances. And this strategy works while undercutting the authority of an elite whose rigid control would lead to the stagnation of science.

Of course the question of how science ought to be organized is different from the question of whether it is so organized. Thus the question of what it would take for science to be rational is different from the question whether science is in fact rational. But now we can guide our empirical studies of the practice of science. And I would be rather optimistic about the answer. The worst historical case against the rationality of science was Feyerabend's. But we may notice that my two requirements of freedom are very similar to his principles of proliferation and tenacity. To those familiar with the literature in the epistemology of science, those requirements are echoes of arguments by Toulmin, Lakatos, and others who have insisted on the generation of alternatives. Insofar as their accounts are based on acute observations of the history of science, the parallel should be expected if science indeed comes close to what it ought to be (in my evolutionary account). The interpretation of results may be different, though. For Toulmin, for example, the generation of alternatives is necessary for his Darwinian analogy. For me it is a requirement that science must meet to increase our adaptability. What seems like anarchy (in Feyerabend's account) under a conception that equates individual rationality with adherence to norms, now seems like appropriate structure, and thus rational, under a social conception born from my evolutionary account.

Incidentally, the need to meet my two requirements partly explains why science is not inevitable. Satisfying our curiosity may be part of nature. But to do it well in a social manner would require a degree of freedom that many cultures would find intolerable. This is not to say that science can only grow in free societies. Tyrannies may permit freedom of cosmological speculation while exerting tight controls elsewhere. But in some other societies freedom of cosmological speculation would lead to the questioning of the religious view of the origin of man and world. Insofar as any such questioning is viewed as a threat to the social cohesion imparted by the society's religion, it will be resisted.

As the problem of justification moves from the individual to the social level, it would be a mistake to suppose that individual irrationality becomes social rationality. The two requirements of freedom must be met by science as a social entity. Different individual scientists can pursue different avenues of investigation in accordance with the promise they see in them. They themselves do not have to be open-minded, nor do they have to entertain alternative hypotheses, let alone develop them. But science as a whole must. Individual rationality still exists but not regarding the choice of theories. It exists regarding the sub-goals a scientist has set for himself; obviously some means will be more effective than others in determining, for example, whether heavy leptons result from certain collisions of high-energy particles. Nevertheless questions of individual scientific behavior and questions about the coordination of such behavior within research groups, and of the interaction between research groups are all important to determine the fine dynamical structures that permit science the meeting of the requirements of freedom. Within an evolutionary context, the psychology and sociology of science can make us aware of tendencies that would lead to the better or worse functioning of science (with respect to attaining its goals). It is not that we accept whatever is done in this or that laboratory or discipline as the way science ought to be. It is rather that we can use such studies to determine which practices tend to, say, do away with the generation of alternatives. As long as these social studies are employed in this fashion, they contribute to a finer determination of what science ought to be like. To add insult to injury (to traditional epistemologists) it turns out that psychology and sociology are by no means irrelevant to questions of rationality and hence of justification.

What I have done can be summarized as follows. I have given an evolutionary account of the genesis of science. From that account I discerned some important features of the nature of science. From those features I concluded that science is a social enterprise and that it has certain advantages to offer. On the basis of these conclusions I was able to determine that rationality would have to be a social property, and also how science should be organized so as to exhibit that property. That is, from my naturalistic study of the origins and development of science I have come to determine what science *ought* to be. This result should not be too surprising. After all, Plato in the *Republic* tried to describe the nature of justice so he could determine its value. Likewise only by understanding the nature of science can we frame properly the question of its rationality and specify the means by which we can answer it.

Some epistemologists may feel uneasy about ascribing rationality when the actions in question are not the result of deliberation. Rationality, they would demand, should result from the careful, certainly conscious, deliberation about the means required to achieve desirable ends. But in my social conception, rationality is divorced from any such conscious deliberation. Their uneasiness should subside when they remember that until recently the very task of philosophy of science was rational *reconstruction*, and thus the rationality of science could not have depended on conscious deliberation by scientists but presumably on standards which, according to Lakatos, scientists applied "sub- or semiconsciously."[11] In any event the means-ends analysis of rationality is compatible with the absence of conscious deliberation. Imagine that on a narrow road a man sees a speeding car about to hit him. He throws himself to the left, where he lands on grass, rather than to the right where he would have landed on rocks. He believed that grass is softer than rock and hence a better place to land in an emergency, though he may have never consciously formulated this belief. Any individual may acquire many beliefs about his environment without ever flashing them across an inner mental screen, let alone deliberating about them (we might eventually tell that he has some such beliefs by his behavior towards that environment). Now, in order to achieve some particular goals the individual may come to act upon those beliefs. The rationality of his actions would depend on whether acting on such beliefs was an appropriate means to his ends, not on whether he consciously deliberated about the whole process.

We must now consider an important consequence of my approach to epistemology. The rationality of science depends on the desirability of ends that it has to offer us. And although our values may lead us to see adaptability as a good thing, it is clear that we also have many other values with which the practice of science may conflict. Why should the values associated with science take precedence over conflicting social and moral values? As we can see, epistemological questions lead naturally into questions of social theory and ethics.

This connection between such distinct areas of philosophy would not be welcome by many epistemologists. It is bad enough to skirt the muddy waters of the rationality of science without having also to trample through the foggy swamps of ethics. Once more, however, we receive guidance from a naturalistic approach.

In this century philosophers have tended to separate nature and value. In one respect this is a very strange attitude, since it appears plausible to suppose that all living things favor some objects over others by the mere fact that being a form of life means engaging in the world in some ways but not in others, taking advantage of some chemical arrangements while avoiding others. More complex forms of life have clear likes and dislikes and display emotions accordingly. The dog values the affectionate word of his master while viewing with suspicion the gait of the stranger coming to the door. If this sounds too anthropomorphic for comfort, let me speak instead of proto-values when we come to humans, it seems clear to me that the very kind of organisms we are predisposes us to find some things important and desirable while others we will be inclined to find of little worth. I do not mean to suggest that our values are hard-wired in our genes, for our development is very flexible, much of it takes place out of the womb, and is thus likely to be heavily influenced by the social environment. For humans and for many other social animals, the social environment becomes the most important factor in their development (even physically, many animals do not mature normally if they are deprived in early life of the typical social interactions of their species).

Even in matters of justification there seems to be at first a plausible connection between nature and morality. To justify one's judgment in ethics, as Peter Singer correctly points out, is to give reasons that take into account not only our own interests, but those of others whom we wish to convince.[12] Realizing that our own interests do not have greater standing just because they are our own (as far as others are concerned) is the key to justification. Values, furthermore, are closely tied to interest in a straightforward manner, for how would values motivate us to action if they are independent of our needs, desires and hopes? This is why appeals to values, via interest, give reasons for action. To fail to give equal weight to the interests of others is to fail to appeal to their interests, and thus to fail to motivate them in the desired ways. But of course, our needs, our desires, and even our hopes are to a large extent influenced by our nature.

The cause of the present separation between nature and value is that philosophers think of the resolution of conflict of values in terms of an inferential model. Statements about nature would be the premises, and from such premises no conclusion containing statements about value may follow. For as Hume presumably showed, facts and values belong to distinct logical categories. Actually Hume did not show any such thing. It just struck him as evident. And part of a philosopher's education is to come to be struck in the same manner. But even if Hume's point is logically unobjectionable, the use that philosophers make of it is not.

Peter Singer, for example, accuses E. O. Wilson of committing the naturalistic fallacy because Wilson believes that biology can give us new values and justify some of the values we hold For Wilson, however, the issue is not whether he can *derive* values from facts, but whether changes in biological knowledge can lead to new values. These two issues may appear similar, but there is a world of difference between them. We all know of cases in which obtaining information about somebody or something leads to a lower or higher evaluation. The standard philosophical response to such examples is that knowledge has not changed our values but has merely made us realize what values we held all along. Thus I run away from the charging bull not because I have suddenly acquired a value that motivates me to run but because the information that the bull is charging me brings into action my long-held value in favor of survival.

The situation, however, is not that simple. Imagine that a black man has grown up hating Whites because he believes that when he was a little child a group of white men drowned his father purely for amusement. This is what the aunts who have raised him have told him; and there is no reason for him to doubt their word, since he has observed firsthand the terrible treatment of blacks in the Southern United States during the 1950's. This man's hatred gets the best of him, and he begins to murder whites. Once he is caught he expresses no remorse — he has very negative values about whites. Some time later he learns from a source he trusts the particulars of his father's death. His father actually was friends with whites. At a picnic with his white friends he fell in the water and drowned. Three of those friends also drowned trying to rescue him. Now, this black murderer may come to change his values precisely as the result of his receiving this information. But Singer denies that any change of values has taken place. According to him, the following sort of inference describes the change in the man's behavior.[13] The man held all along that anyone who would drown another for amusement must be evil (and so must be other members of his race). This proposition which embodies his real value serves as the major premise in an argument. The minor premise is that whites drowned his father for amusement. The conclusion is that whites are evil. This conclusion, according to Singer, does not express a basic value. The black murderer gives up this conclusion, then, when he finds that the minor premise in the argument is false.

I find this inferential account of value very implausible; among other reasons because Singer and I agree that values are the sorts of things that motivate us to action; and the only value expressed in this argument, if Singer is correct, can hardly prompt murder. Moreover, if it did, it seems that anyone else who accepts the two premises should also be inclined to murder. Some psychological supplement may perhaps be offered to strengthen Singer's position, but I think that a much more straightforward account is at hand.

The black murderer has a posture of extreme hostility towards the world, and particularly towards whites. This posture comes from having come to think of himself as a victim, as a man who has been robbed of the love of his father by the sadistic whims of white society. His life has been destroyed from the beginning and having thus nothing to lose he asserts himself by brutal revenge. Learning how his father really died does away with the reason for his posture. He has been a victim indeed: of a lie. New knowledge changes his view of himself and his relation to the world. That is why he is likely to change his negative values toward whites: Those values do not fit in with his new perspective. This analysis of my example suggests a causal model not only of the origin but of the justification of value.

To see the point clearly let me consider a second example. The North American Indians presumably held a high regard for the environment as a cardinal value. Nevertheless, I imagine that they probably did not arrive in North America with that value firmly established in their culture (indeed many species of big animals became extinct as the result of the human invasion). As they came to adapt to the new environment, a form of that value arose and developed slowly in interaction with the Indians' ideas about their new world and their place in it. After a long time a whole network of beliefs about the world and about themselves came to cohere with a set of appropriate values. Given their beliefs, those values were essential to their well-being and probably to their survival.

I suggest that this causal model applies quite often in individual and social cases. To explain the origin and development of a value (the genetic fallacy again!) may well amount to explaining why it coheres with a network of beliefs about the world. If that network is deemed acceptable we may thus find the value justified. If the relevant views of the world have been undermined, however, the values may come into question; for they may now fail to cohere with the emerging alternative views of the world. An example of the first situation we find in the case of the American Indians. An example of the second, in the case of the black murderer.

Values may also be criticized by being made to conflict with other values in the network. Our set of values is always potentially inconsistent (in places it may also be formally inconsistent). What I mean is that values arise in connection with our dealings with certain areas of experience. We are complex beings, however, and other circumstances lead to the development of other values. Conflict between these two subsets of values may never arise unless new circumstances bring about a tension between the two ways of behaving towards the world. We may, however, criticize some values or defend others by either pointing to an already existing inconsistency or by making a plausible case that a potential inconsistency will become actual and it would be advisable to make a choice now. The reason for choosing one way rather than another is that the arrangement of the network consistent with the first is on the whole deemed more likely to produce fruitful behavior.

The philosopher in search of justification must then become a rhetorician in the best sense of the word. His aims to persuade are satisfied only by an account that maximizes the interaction between human nature and the world. A justification that appeals only to a narrow group of desires and inclinations can be easily challenged by pointing to unwelcome disruptions elsewhere in the network of beliefs and values. A decision to prefer a particular arrangement of the network is based on a valuation of that arrangement, of course, and that valuation is itself open to criticism. My references to human nature, and the values thereof, should not be interpreted to mean that there is a fixed, ultimate court of appeal in the resolution of conflicts of values. Seemingly basic tendencies give rise to values in a variety of ways, and those values may also be challenged by arguing for a further rearrangement of the network of values and views of the world. But the freedom to challenge any such arrangement does not put all arrangements on a par. The situation is the same here as it was in the epistemology of science: the hard work still is ahead of us.

In the resolution of conflicts of values, the task of justification is thus very complex. To carry it out successfully, the philosopher is well advised to understand how human nature lays out the pathways and limits within which justification may operate. Let me explain. It may be thought that biological, and perhaps social, factors influence morality only at what we may call the level of conduct. At a deeper, theoretical level we find the basic principles of justification (e.g., the principle of utility). At this level, however, biology has nothing to tell us (nor psychology, nor anthropology, nor history). There is no biological discovery that could possibly count against the principle of utility. Nevertheless, once again, a little biology takes us a long way. Indeed, it takes us far enough to put clamps on the principles of impartiality and utility.

The principle of the impartial consideration of the interests of all concerned would lead us to realize that our individual interests do not count more than the interests of others in our group just because they are ours. Similar considerations will apply to the interests of our group with respect to those of other groups, of our society with respect to those of other societies, and so on. In this manner, Singer points out, we expand the circle of ethics. Nevertheless Singer recognizes some problems with the universal expansion of the circle. The main problem is that it may do too much violence to human nature by ignoring the special obligations that we feel. We cannot be expected to act upon a morality that would have us treat our own children no better than we would treat strangers. This would be a morality suitable for "disembodied spirits" only, as Wilson has remarked. If human nature, in spite of its plasticity, strongly prefers to run down some streams rather than others, it would do us well to understand the lay of the land. Accordingly we restrict the principle. We now wish to say that our need to act in a special way towards our own children does not count more than the need the next fellow feels to act in a special way towards his own children, just because it is our need. With this restriction upon the principle of impartiality, we are now ready to understand the influence of biology upon other basic principles of morality, or perhaps, I should say, of ethics.

The best way I know to apply the principle of impartiality is to use John Rawls's veil of ignorance, both to choose a fair society and to decide on the right individual action to take in case that the general moral structure does not provide sufficient guidance.[14] Now in choosing the basic social and moral principles by which a society is to organize itself, agents who would not know what position they are to occupy in that society would find it in their best interest to prefer a fair arrangement, so as not to be unduly victimized and so as to maximize their lot regardless of their actual position. Behind such a veil of ignorance, thus, a sensible person would choose a social arrangement that would permit the maximum amount of individual freedom compatible with equal freedom for others. A sensible person, however, would reject the principle of utility. The reason is that the principle of utility tells us that the right action is that which produces the largest balance of happiness over unhappiness; but the principle is neutral as to whose happiness is enhanced. Thus utilitarianism would permit a great deal of unhappiness for any one person in order to pay for the happiness of others. According to Rawls, impartial consideration of interests will be assured because each agent will be trying to maximize his chances under the special conditions of the veil of ignorance. But no agent is going to agree to an arrangement under which his happiness may be completely ruled out from the start. This is why the principle of utility would be rejected. As Rawls has pointed out, it cuts too much against the grain of human nature. Singer has argued that if the principle were rejected on these grounds we would be implying that it would make people too unhappy to use it. But in that case we would still use a consequentialist, indeed utilitarian, criterion for what we ought to do.[15] His argument misses Rawls's point, though. The point is that, in trying to secure impartiality, it is not sensible to employ the principle of utility.

This conclusion is satisfying enough in that it gives a most crucial role to naturalistic considerations. But naturalism goes even further than that. Normally when we talk about the relationship between biology and ethics we deal with the same human biology and ethics. Let us consider what might happen if we vary one of the factors. Imagine the sort of morality that rational ants might have. By rational ants I simply mean that they are roughly as intelligent as humans are. As we know, the biology of ants and other social insects brings about systems of social organization and behavior very different from ours. We may realize also that, as Singer himself puts it, the origin of morality is to be found in the social character of our ancestors. To remain social an organism must, at least in some instances, restrain its behavior toward others and make some effort in their behalf as well. It is at least plausible to suppose that the different social behavior of the ants would then lead to a different morality. In order to make a firmer determination, let us then place a rational ant before Rawls's veil of ignorance. A rational ant, if Wilson is correct, would abhor the prospect of individual freedom, since it would cut too much against the grain of its nature to attempt to maximize its own happiness. Thus rational ants would not choose Rawls's principle of individual freedom as a basic principle of morality. On the one hand, rational ants would find the principle of utility quite congenial. In crucial respects, then, their morality would be the reverse of ours.

The moral of this story is that naturalism indeed must pervade the task of justification, even when we make distinctions between surface and deep levels. We must keep in mind, incidentally, that in my account considerations about impartiality and the like are to be subsumed under the more general schema for criticising and rearranging the network of values and views of the world. By this I mean that in resolving a conflict of values it is not always enough to give equal weight to the interests of all concerned with respect to the issue in question. In this specific sense of interests, we face a whole network of relationships with other interests, beliefs, and so on. In some cases it may be sufficient to propose parity between the values of different agents; but in other cases we forego parity and aim instead to have the other agent give up or modify some interest through a rearrangement of his network (which our criticism of some of his beliefs may make urgent). It all depends on whether we share the same interests, as in the case of our wanting to give special treatment to our own children, or not.

Both in epistemology of science and in ethics the distinctions and fallacies of the philosophers' invention have become barriers to philosophical understanding. Even if philosophy deals with norms we may profit from a good dose of naturalism. For in understanding how nature made us normative, we gain the proper background against which genuine philosophical questions arise. Our philosophy thus has a chance when we view human thought within the context of natural history.

Notes

1. C. G. Hempel, "Scientific Rationality: Normative vs. Descriptive Construals," in *Wittgenstein, the Vienna circle and Critical Rationalism*, Proceedings of the 3rd International Wittgenstein Symposium (1978), p. 291.
2. Particularly T. S. Kuhn, *The Structure of Scientific Revolutions* (Chicago: University of Chicago Press, 1970); and P. K. Feyerabend, *Against Method* (London: NLB, 1975).
3. Hempel, "Scientific Rationality," *op. cit.*, pp. 298-99.
4. Feyerabend, *Against Method, op. cit., Science in a Free Society* (London: NLB, 1978); I. Lakatos, "The Problem of Appraising Scientific Theories," in his *Mathematics, Science and Epistemology*, J. Worral and G. Currie, eds. (Cambridge: Cambridge University Press), pp. 107-20.
5. For a fuller account of this problem and its historical background, see Ch. 7.
6. Particularly K. R. Popper, *Objective Knowledge* (Oxford: Oxford University Press, 1972) and S. Toulmin, *Human Understanding*, vol. I (Princeton, NJ: Princeton University Press, 1972).
7. W. V. Quine, "Epistemology Naturalized," in *Ontological Relativity and Other Essays*, (New York: Columbia University Press, 1969).
8. J. Piaget, The Psychology of Intelligence, (Totowa, NJ: Littlefield, Adams & Co., 1972).
9. For a fuller account of this naturalistic approach see my *Radical Knowledge: A Philosophical Inquiry into the Nature and Limits of Knowledge*, (Indianapolis, IN: Hackett, 1981).
10. I begin the task in my *Radical Knowledge*, ibid., and treat the matter in great detail in *The Dimming of Starlight*, still in preparation.
11. Lakatos, "Appraising Theories," *op. cit.*, p. 110.
12. P. Singer, *The Expanding Circle: Ethics and Sociobiology*, (New York: Farrar, Strauss, Giroux, 1981).
13. P. Singer, "The Expanding Circle: A Reply to Munevar," *Explorations in Knowledge*, vol. IV, no. 1 (1987), pp. 51-54. Several of the examples I give here first appeared in my critique of Singer's *The Expanding Circle, Ibid.*, pp. 43-50.
14. J. Rawls, *A Theory of Justice* (Cambridge, MA: Harvard University Press, 1971).
15. Singer, "Reply to Munevar," *op. cit.*, p. 53.

11 The Morality of Rational Ants

Introduction

David Hume once wrote, 'The mind of man is so formed by nature that, upon the appearance of certain characters, dispositions, and actions, it immediately feels the sentiment of approbation or blame'.[1] In this, the Darwinian era, it seems intuitive to say, with Michael Bradie, that 'The full range of human endowments that make up human nature — physical, intellectual, and moral — are . . . the result of eons of evolutionary development'.[2] Presumably, however, creatures who have taken a different evolutionary path will be compelled by different complexes of moral emotions and may thus exhibit a different morality. This point is, of course, the analog of the evolutionary relativism of Part I, and Bradie has recently expressed some sympathy towards it by defending the notion that Darwinism seems to commit us to an interspecific relativism, at least where morality is concerned. As he says, '. . . we can safely conjecture that any alien species that has evolved to the point of having moral sensibilities is liable to have sensibilities very different from our own'.[3]

In this chapter I will examine the scientific and philosophical developments that make it plausible to reason in this manner, and I will also determine the philosophical significance of such developments. The outcome will be a recommendation to do ethics in a new way.

The Evolutionary Origin of Morality

To help me accomplish my task, I will make extensive use of Peter Singer's discussion and criticism of the relevance of biology to ethics as they appear in his challenging book *The Expanding Circle: Ethics and Sociobiology* (*EC*).[4] Singer begins with a sympathetic account of sociobiological claims about the origin of morality. Social animals, he points out, restrain their behavior toward each other and do things for each other. Since our ancestors were clearly social animals, 'we can also be sure that we restrained our behavior toward our fellows before we were rational human beings' (*EC*, p. 4). In this he follows not only the sociobiologist E.O. Wilson,[5] but ultimately Darwin himself, who wrote that 'The foundation [of moral qualities] lies in the social instincts, including under this term the family ties'.[6] The origin of morality is, according to Singer and Wilson, not to be found in rational social contracts but in our evolutionary past as expressed in our genes.

The connection between social behavior and morality is established in large part by observing the behavior of social animals. Since it would be philosophically difficult to prove that animals behave morally, the sociobiologist points instead to the seemingly altruistic behavior of social animals and then tries to explain how morality could have arisen from that altruism. Many animals do indeed act so as to benefit others at some cost to themselves: blackbirds warn their fellows of hawks flying overhead, even though by doing so they call the hawk's attention to themselves; African wild dogs risk their lives against cheetahs to save a pup; other species share food or help injured animals to survive. As Darwin pointed out a century earlier, 'Animals endowed with the social instincts take pleasure in one another's company, warn one another of danger, defend and aid one another in many ways'.[7] The problem for sociobiology is to explain how altruistic genes could spread through a population when they seem to make the individual altruist less fit. The solution is presumably that natural selection takes place at the level of genes and that complexes of genes that lead to altruism are favored in a variety of ways. The first obvious way is kin altruism. The bits of matter which make up the genes of an individual will perish with him, but their type can be passed on by reproduction. In this sense the genes survive by being passed on. It is to the genes' advantage then to produce an individual that will act so as to benefit other individuals that have the same genes – in some cases with his life, for by his death his kin may live on, thus ensuring the survival of the genes in question. The natural selection of such genes, kin selection, determines the (genetic) inclusive fitness of an individual, which thus includes not only the reproductive success of that individual but also the reproductive success of its relatives.

Reciprocal altruism expands the circle of individuals which an animal benefits at some cost to itself. It is not clear, however, how the appropriate genes may spread through a population. To explain that, Singer resorts to some ingenious tales in which an isolated population influenced by kin selection favors reciprocity in several forms (monkeys grooming each other, for example). If perchance such a group can come to dominate the population, the reciprocating genes may then come to predominate in the population as a whole. It seems to me that tales of this sort are made more plausible if we realize that the genes that lead to certain tendencies (e.g., to cooperate with our kin) may be expressed through phenotypic mechanisms which only loosely favor kin (e.g., to groom those who are around us), and that these mechanisms may become advantageous in their own right and may thus extend beyond kin. Just to be clear, the genes that would spread throughout the population are not genes that extend kin altruism to the whole species, for as Darwin pointed out the social instincts are not likely 'to extend to the whole species, but only to those of the same community.'[8] In any event, this extension of kin selection amounts to a form of group selection, which is still a matter of debate. I cannot enter that debate here, but I would like to point out that not all forms of group selection are equally objectionable. It is true that some traits that favor the group may actually work to the detriment of some individuals, but in the present case we are considering traits that by enhancing the survival of the group (e.g., by increasing its social stability) also increase the inclusive fitness of the individual members of the group.

The extension of these ideas to human ethics seems implausible to many who are impressed by the great diversity of moral customs among human beings; diversity presumably brought about by cultural influences. But as Singer remarks, 'there are common elements underlying this diversity. Moreover, some of these common elements are so closely parallel to the forms of altruism observable in other social animals that they render implausible attempts to deny that human ethics has its origin in evolved patterns of behavior among social animals' (*EC*, p. 29). For example, the strength of kin altruism among humans is illustrated by the failure of attempts to do away with the family so as to channel the loyalty of people toward their community (as Plato recommended in *The Republic*, and the Israelites and Russians tried to put into effect). Reciprocal altruism, Singer points out, is also prevalent in human societies. Some critics have charged that since reciprocity involves self-interest it is not compatible with altruism, and thus the very expression 'reciprocal altruism' involves a conceptual confusion. In genuine altruism an agent benefits others at some cost to himself, and in doing so he is only motivated by the desire to benefit those others. But Singer points out that this is perfectly compatible with the notion that in fact genuine altruists may be more likely to pass on their genes than self-interested individuals. As Singer argues, 'the existence of real-life Prisoner's Dilemma situations puts egoists at a disadvantage where cooperation is advantageous [as it often would be for social animals]. In these situations two genuine altruists will do better than two egoists, and a single egoist will not do as well as an altruist if her egoism is apparent to others' (*EC*, p. 128; the Prisoner Dilemma cases are discussed in pp. 45-49).

This point is often overlooked, even by some sociobiologists. Some of them argue against the possibility of universal altruism (e.g., Richard Dawkins and Garret Hardin) on the grounds that natural selection evolves only organisms that obey the mandates of the 'selfish genes.' Individuals are 'selfish' insofar as they act to preserve their inclusive fitness. But being fit is understood in terms of helping one's genes to survive. Thus behavior that increases inclusive fitness, which would be 'selfish' according to this sociobiological sense of the term, is not really selfish. For such behavior may well be motivated, at the individual level, by the desire to contribute to the welfare of others (a mother's love for her children may be determined by her 'selfish genes': but when she sacrifices her life so her children may live, she still acts out of love, not because she has calculated that her death would increase the chances of her genes). If Singer were not right about this, sociobiology could not explain the emergence of the moral emotions, as I believe it does.

I think the clarity of this position undermines the motivation for some recent evolutionary treatments of morality. For example, Richard Alexander conflates an individual's interests with his inclusive fitness[9], a move which most philosophers should find very unintuitive and difficult to accept, appeals to evolutionary thinking notwithstanding. It seems to me, though, that I can think of my interests in the traditional way, while using sociobiology to explain why it is that I have these impulses to sacrifice my interests for someone else's welfare. Indeed, being able to do this was one of the principal merits of sociobiology's application to ethics.

It also seems to me possible that some animals may turn out to be moral, although let me emphasize that I advance this merely as a possibility. Darwin himself restricted morality to beings capable of reason and the power of deliberation. But reason and deliberation, as Bradie points out, are not 'saltative' qualities: The difference between humans and other animals is one of degree, not of kind. There is no need, then, for distinctions such as that invoked by Robert Richards between motives and intentions. Morality requires intention, but even though animals can act out of altruistic motives, they do not perform those acts intentionally.[10] Apart from the fact that this move runs contrary to the non-saltative spirit of Darwinism, it involves the perplexing commitment to non-intentional motives. Perhaps such a commitment may seem less perplexing if we take altruistic impulses (which could be instinctual, or at any rate unconscious) to be motives. But I think that Singer's account can keep us from having to ascribe motives to the instinctual actions of an ordinary ant.

If some animals do turn out to be moral (perhaps chimpanzees or dolphins), we should keep in mind that their inclusive fitness, as well as their capacity for reciprocal altruism, are likely to be expressed in forms quite different from ours. And we can also imagine that rational ants would be even more different from us. The different moralities that would result may be no less 'valid' for them than ours is for us. We will see below what consequences this point has for ethics.

The Relevance of Biology to Ethics

Singer's sympathetic account of the sociobiological explanation of the origin of morality does not lead him to agree with E. O. Wilson's proclamation that 'the time has come for ethics to be removed temporarily from the hands of the philosophers and biologicized' (*EC*, p. 55). Singer offers serious objections to the three main ways by which Wilson envisages the takeover bid by biology. According to the first, biology should make us aware that certain moral decisions produce bad consequences because they go against human nature (the biological tendencies that constitute part of human nature can be fought and blunted, of course, but sometimes the cost may be so high as to be unacceptable, as in the attempts to do away with the family). This new knowledge of biological consequences is welcome, Singer agrees, but it does not affect moral theory. If a philosopher is a consequentialist, he will already think that the morally correct action is that which produces the best consequences. Biology may thus help him with specific moral judgements, but it will do nothing to alter his consequentialist philosophy. If on the other hand the moral philosopher holds that consequences do not affect the moral character of an action, a new list of consequences is not going to alter his philosophy either. 'There is', Singer concludes, 'no justification here for [...] fashioning a biology of ethics which will do away with the need for ethical philosophers' (*Ec*, p. 68).

The second way in which Wilson thinks that biology can transform ethics succeeds all too well: biology may undermine our ethical beliefs. According to Singer, 'almost all the thinking we do about ethics involves connecting one ethical judgement to another, more fundamental one'. (*EC*, p. 70). And even some moral philosophers, he points out, 'explicitly say that philosophy can do no more than systematize our moral intuitions' (*EC*, p. 70). It is here that biology can have a most devastating effect, for it can show us that 'what we take as untouchable moral intuition may be no more than a relic of our evolutionary history' (*EC*, p. 70). The lesson that we ought to draw is that 'discovering biological origins for our intuitions should make us skeptical about thinking about them as self-evident moral axioms' (*EC*, p. 70). Since discovering the cultural origins of the rest of our moral intuitions has the same effect, if Wilson is right we run the risk of debunking all of ethics. It is important to realize, then, that 'the effect of a demonstration that some form of behavior has a biological basis can be the opposite of what those who try to deduce ethical principles from biology usually claim' (*EC*, p. 71).

Here moral philosophy does face a problem. Where does this debunking of ethics by biological and cultural explanations stop? According to Singer, Wilson's answer might be that we should retain those principles that 'will be shown to be biological adaptations which remain well suited to the contemporary human situation' (*EC*, p. 72). This implies, however, that biology can serve as the justification for some of the values that we already hold, and perhaps, as Wilson claims, suggest new values. This is Wilson's third way of biologizing ethics. But is it not an instance of the naturalistic fallacy? How can biology bridge the gap between facts and values?

Singer does not believe that Wilson can manage to perform that trick. His suggestion is that we can stop the debunking of ethics only if we realize that ethics has a rational component. According to Singer, 'Ethics starts with social animals prompted by their genes to help, and to refrain from injuring selected other animals. On this basis we must now superimpose the capacity to reason' (*EC*, p. 91). The difference between the proto-ethics of social animals and the full-blown ethics of humans is that the latter involve moral judgements. But to judge, 'beings have to be capable of thinking and of defending the judgements they make' (*EC*, p. 91). And here comes the critical contribution of reason: 'in making ethical decisions I am trying to make decisions that can be defended to others. This requires me to take a perspective from which my own interests count no more, simply because they are my own, than the similar interests of others' (*EC*, p. 109).

This principle of the impartial consideration of the interests of all concerned enables Singer to reply to the objection of those who think that ethics has nothing to do with interests because whereas 'the interests of humans – and other animals – are short-lived and parochial, the laws of ethics are eternal and universal' (*EC*, p. 105). If there is something eternal and universal about ethics, Singer argues, it is the principle of reasoning which would permit any rational, social being, 'whether on Earth or in some remote galaxy' to conclude that his 'own interests are one among many sets of interests, no more important than the similar interests of others' (*EC*, p. 106). This universal element of ethics, however, 'exists only as a framework into which the deliberations of rational creatures with preferences fit' (*EC*, p. 106). Without the existence of beings with interests, the framework would have no application – since the requirement that we treat all interests equally would be empty.

Eternal moral truths and 'objective' values existing independently of living creatures would be very mysterious entities. To be aware of them, J. L. Mackie says, we would need 'some special faculty of moral perception or intuition, utterly different from our ordinary ways of knowing anything else'

(*EC*, p. 107). But even if they existed, it would be difficult to see how they could serve as the basis for moral action. For, as Singer adds, 'values are inherently practical; to value something is to regard oneself as having a reason for promoting it. How can there be something in the universe, existing entirely independently of us and our aims, desires, and interests, which provides us with reasons for acting in certain ways?' (*EC*, p.107). Until a plausible account of such truths has been given, Singer suggests, we should 'cling to the simpler idea that ethics evolved out of our social interests and our capacity to reason' (*EC*, p. 111).

Subjectivist objections do not present a greater obstacle than absolutist views. If to defend a moral rule is merely to express a subjective preference for it, then, as Singer points out, impartially 'this preference should not [...] count for any more than any other preferences of similar strength' (*EC*, p. 108). We must still face the task of 'impartially adjudicating a conflict of preferences' (*EC*, p.109).

Singer thus accounts for the origin and nature of ethics. As he says, 'the idea of a disinterested defense of one's conduct emerges because of the social nature of human beings and the requirements of group living, but in the thought of reasoning beings, it takes on a logic of its own which leads to its extension beyond the bounds of the group' (*EC*, p.114). for to claim that what I wish to do is right I must give some reason other than the fact that my action would benefit me, my kin, my village, or my society.

Singer then deals with two important objections to his view. The first comes from Hume's remark that 'reason is and ought to be the slave of the passions'. If Hume is right, reason cannot move us to act beyond the bounds of kin, reciprocal, and group altruism. But Hume is not right, Singer argues. Even if reason is a tool, 'tools have a way of influencing the purpose for which they are used [...] In the case of ethical reasoning, we begin to reason impartially in order to justify our conduct to others, and then discover that we prefer to act in accordance with the conclusions of impartial reason' (*EC*, p. 142). Reason is also part of our biological nature; as such it also may place demands on our feelings and motivations. In particular, going against the consistency required by the principle of impartiality may create in us a 'cognitive dissonance' (*EC*, p. 143). To avoid the discomfort that inconsistency creates in us, we sometimes are inclined to behave consistently with our belief – and on occasion with our public pronouncements. Reason may thus generate its own passions.

But if reason can thus expand the circle of ethics, we must consider whether a morality based on the principle of impartiality alone would not be, in Wilson's words, 'an ideal state for disembodied spirits'. Singer acknowledges the point: 'we may attempt to foster tendencies that are desirable from an impartial point of view and to curtail the effects of those that are not; but we cannot pretend that human nature is so fluid that moral educators can make it flow wherever they wish' (*EC*, p. 155). A rational ethical code therefore, 'must also make use of existing tendencies in human nature' (*EC*, p. 155). In practice, this means that if we want to induce people to donate money to reduce hunger in the world, we do not appeal to universal altruism but try instead to make personal appeals. Humans are not inclined to give money to anonymous beneficiaries, but when we become foster parents of poor children we draw on our kin altruism, if only by extension.

The rules of a morality that people have a chance to live up to will limit our obligations, make them more personal, and be easy to teach and to apply as a first approximation. 'Do not kill innocent human beings', is one such rule. 'Preserve innocent human lives', on the other hand, would often require us 'to give up everything and work full-time to save the lives of others' (*EC*, p. 160). Saints may bear that burden, but we simply cannot expect everyone to act accordingly. 'We must begin to design our culture so that it encourages broader concerns without frustrating important and relatively permanent human desires' (*EC*, p. 170). To do this, however, we need the knowledge of human nature that sociobiology can provide. Thus it seems that even if the biologist cannot take over ethics, the moral philosopher will still need his advice as our reason tries to master our genes.

The rules of morality that other intelligent creatures would find advisable are, then, likely to be different from ours to the extend that their natures provide them with tendencies different from ours. Their special obligations, for example, will correspond to the manner in which the specific range of kin and reciprocal altruisms are expressed. A rational worker ant may be expected to run into a burning building to save her queen, but not to risk her life for a distinguished drone. To forestall unnecessary controversy, by the way, let me say that I am speaking of the nature of a species in a populational, not in an essentialist way.

The issues just discussed point to some possible limitations of Singer's appeal to the impartial consideration of interests. The most important concerns the extent to which interests and morality coincide. We will see below that in Rawls' Veil of Ignorance the interests in play are not the individual's actual interests but rather what those interests might be in a hypothetical situation. And when Singer himself draws back from perfect impartiality he is prompted by the recognition of strong biological tendencies that result from inclusive fitness and other factors that go beyond individual interests, i.e., beyond the kind of interests likely to fall within the purview of reason. Beneficence, to use the favored 18th Century term, may also be wired into our genes (although not exceedingly hard, I presume).

No longer directed by reason alone, how smoothly will the expansion of the moral circle proceed?

Like many other good books, *The Expanding Circle* opens up new approaches to the subject-matter, and presents the opportunity for questioning established views. This opportunity arises not only where Singer argues successfully for his point of view on crucial matters but where, in my opinion, he does not. Before discussing the latter, I feel obliged to quibble with Singer on a couple of points which he finds important although they do not affect significantly the main argument in the book. The first is Singer's insistence throughout that the principle of impartiality requires a consequentialist theory of ethics. As we will see below, for example, Rawls' Veil of Ignorance is a perfectly adequate way of satisfying the requirement of the impartial consideration of the interests of all concerned.

My second quibble is with Singers claim that the circle of ethics should be expanded to many animals on the grounds that, since they can feel pleasure and pain, they have interests also. Plants have no interests because they are not sentient, and oysters are a borderline case. But Singer has not shown that we should extend the circle of ethics to all sentient creatures. He has shown that when we have to justify our conduct to others we must take their interests into account impartially, otherwise our reasons are not likely to convince them. But surely there is no question of justifying our conduct to cows or giving reasons that chickens will find convincing. There may be arguments that show that we have moral obligations toward animals, but those are not the arguments that Singer gives in his book.

I would like to discuss now the two crucial issues where Singer parts company with the sociobiologists. Since, in his disagreement, Singer upholds the philosophical tradition against outsiders, a critical examination of his position may tell us much, not only about the connection between biology and ethics, but about the nature of ethics itself.

Biology Defeats the Naturalistic Fallacy

To begin with, let us recall that Singer chides Wilson for arguing that biology may suggest new values and help us to determine which values to keep. Singer rushed to point out, as most philosophers would, that we cannot derive values from facts. But is Singer's objection at all relevant? The naturalistic fallacy is one of those 'discoveries' that make philosophers feel that they understand something other people do not. But I think that it is an intellectual red herring. For other people, Wilson included, the issue is whether knowledge can make us change our values, not whether we can logically derive values from facts.

The difference between Wilson's perception of the issue and that of the philosophers is quite pronounced. I think that Wilson's perception is right, and with the reader's indulgence, I will sketch by means of two rather lengthy examples, a plausible case against the philosophical tradition. My first example goes as follows. A black man hates white people so much he has taken to murdering them. His hate began when he was a child and a group of white men took his father to a lake and drowned him for their amusement. No one even bothered to bring charges against them — in those days it would have been useless — but now this black man is exacting a heavy price from the white race. He is caught and sentenced to jail, but he does not feel sorry for his actions. If he had a chance he would gladly commit the murders all over again. Without a doubt he has some very negative values about whites; and those values give him strong reasons for action.

While this murderer is awaiting transfer to a permanent jail, he meets an old black man who is awaiting trial on a vagrancy charge. After much conversation they are both pleasantly surprised to realize that the old man knew the younger's father well. Many reminiscences of happy times later, the old man mentions that he was present at the unfortunate death in the lake. The younger man presses him for details. And then he finds to his astonishment that his father was not murdered by white people. On the contrary, he was very good friends with them. Indeed, he was at a picnic in the lake with his white friends when he fell of the boat. Three of those friends drowned trying to save him!

All of the black murderer's information about his father's death had come from the aunts who brought him up (his mother had died much earlier). We can only speculate that his aunts so much disliked white people, so much opposed his father's befriending them, and so much expected the worst to befall him, that they did one of those hateful things that people sometimes do: They told the little orphan a terrible lie. In any event, it is not implausible that this new knowledge may have a most profound effect on this murderer's values, that he may stop hating whites so intensely and come to feel great sorrow for his crimes.

In his novel, The Tent of Miracles, the great Brazilian writer Jorge Amado describes the virulent prejudice against blacks and mulattos at the turn of the century in Brazil. Many distinguished scholars at the School of Medicine in Bahia wrote treatises on the inferiority of blacks and the abomination against nature and civilization inherent in the mixing of the races. Some proposed sending the mulattos to some big reservation in the middle of the Amazon jungle until a satisfactory final solution could be thought out (the blacks could be kept for hard labor, but the mulattos were congenitally lazy and of bad moral character). The hero, a mulatto, did much to stop the racist propaganda when he published a genealogy of the best families in Bahia, in which he showed that practically everybody in town, including the most prominent racists, was of mixed blood. When the knowledge finally sank in, it had the most devastating effect. It is difficult to go on holding the same obnoxious values when you realize that they give you reasons for acting against yourself.

Although in this second example it is rather clear why new knowledge would lead to a change in values, I do not mean to suggest that the same explanation applies to the first example, nor that an explanation may be provided in every case. A discussion of that subject would take us too far afield. Now, when philosophers are confronted with examples in which 'facts' give directions for acting, they are quick to point out that some background values have come into operation. Singer is no exception. As he says, 'the fact that the bull is charging does not, by itself, entail the recommendation: "Run!" It is only against the background of my presumed desire to live that the recommendation follows' (*EC*, p. 79). But far from being a good reply, this point only illustrates the irrelevance of Singer's objection against Wilson.

It is presumably only in a fanciful sense of the term that we would say that we had acquired the 'value' to run away from the bull. We can describe what we do and explain it without adducing any such new value. But the situation is completely different in my two examples. We would be using the term even more fancifully if we were to deny that the white racists from Bahia, or the black murderer, have changed their values. Nor can we say, as Singer does against Wilson, that the facts have not given us new values, that they have only helped us realize that we held certain values (presumably all along). For the black murderer is almost certainly giving up values he held strongly and replacing them with others he did not have. And his transformation comes about because the new knowledge makes the old values appear untenable.

Singer is not persuaded by my examples.[11] He suggests that the black murderer did not have 'whites are bad' as a fundamental value, but instead held something to the effect that people who would drown a man for amusement are bad, and so are all their race. This value would function as a major premise in an argument, of which the second premise is that white men drowned his father for amusement. When this minor premise is found to be false, he withdraws his conclusion that whites are bad.

Nevertheless, this reconstruction is rather far-fetched, and it is difficult to see how the value encapsulated in the (doubtful) major premise could lead one to action, indeed to extreme action. In the previous chapter I suggested a more plausible explanation: His aunts' lie led the young boy to adopt a posture of extreme hostility towards a world that so cruelly robbed him of his father. As he grew, so did his rage. What he thinks of the world determines in part what he thinks of himself. When he learns the true story, that hostile, violent posture no longer makes sense. He can see now the damage he has caused for what it is, and he feels regret. This is not the straightforward inferential process so dear to philosophers, but a complex arrangement of beliefs and values that, I hope the reader can see, *does* lead to action. A change in knowledge thus may bring about a change in values without the hindrance of having to derive an 'ought' from an 'is'.

As for the second example, Singer suggests that the racists' values were not ethical but only self-interested, for they were not prepared to universalize their judgement (otherwise they would have concluded, in all consistency, that they should consider themselves inferior once they learned that they were of mixed race also). It seems, however, that, if anything, this example puts into practice Singer's recipe for expanding the circle. Amado's hero appeals to the interests of the racists, and they end up taking the interests of *all concerned* into account (which include their interests as well). Indeed, this example seems to provide a means by which we can resolve many conflicts of values: Show the other side how their values are connected with views of the world, or with other values, that on reflection they should find untenable. Incidentally, I must confess the cynical suspicion that if consistency, let alone the commitment to universalize, were requirements for holding ethical values, most of us would be bereft of ethical values altogether. Methinks a confusion lurks here between a standard for the criticism of ethical values and a criterion for having them.

Is it not the case, however, that the new knowledge in my examples had to work hand in hand with some other values for the transformation to take place? That may well be, although it is difficult to determine just what the other values might be. But this admission does not help the standard philosophical view of the matter. Wilson's position is that biology can give us new values. This Singer can no longer deny. What he can perhaps deny is that biology in an intellectual and emotional vacuum can give us new values. But why would Wilson even bother to make such a claim? For someone who is well aware of the inclinations that we receive from biology and culture it would be preposterous. On the contrary, it is against that background of inclinations (of values) that he attempts to select some and add others. Building on kin selection, he suggests that with proper attention to the matter we may begin to cherish the human gene pool as a cardinal value because there is a sense in which we are related to mankind as a whole (for our DNA was once distributed among millions of people and will once again be dispersed to many more millions). Plato tried to accomplish the same sort of thing when he suggested in the first part of the Myth of the Metals that all men in the city are brothers. Like Singer, I do not think that Wilson quite succeeds. But the failure of this attempt does not indicate a failure of method. Singer himself uses the very same method when he proposes that universal altruism may be a successful moral standard if channeled through kin and other biological forms of altruism (e.g., having people become foster parents for poor children). In other attempts Wilson is more likely to succeed, for example when he proposes that the diversity of the gene pool should be a value.

Whether we acquire such a value depends on many factors. But that it could become a cardinal value should not be blocked by fears of the naturalistic fallacy. It could be said that some North American tribes held respect for the environment as a cardinal value. It is quite likely, however, that such respect did not start out as a cardinal value. Instead we may imagine that respect for the environment was so needed for the Indians' survival, that it eventually became an integral part of their way of life. That is, it eventually became a cardinal value. What Wilson is attempting to do is not different in principle. The logical positivists, of all philosophers, already pointed out that values can be criticized by establishing the falsity of some of the facts with which they are associated. But the interplay between facts and values goes way beyond that. And a good thing too. Otherwise I suspect that human culture would be a lot more static than it is, and than it ought to be.

Moreover, when philosophers accuse others of committing the naturalistic fallacy we should wonder whether the culprit's position has been unfairly distorted. Take the most notorious example we can find in the literature: the Social Darwinists' alleged claim that being in agreement with evolution was right while impeding evolution was wrong. This claim was, however, part of a larger argument, namely that evolution worked to perfect living things, so that through its action we moved on to a better stage of the world. If this were true, to facilitate evolution would help bring about a better world, while to impede evolution would have the opposite effect. Thus the original claim would follow. This view of evolution is neither true nor Darwinian, but it does not involve an instance of the naturalistic fallacy. The 'values' were built into the larger argument. Only by taking the claim out of context did it look as if the moral reasoning was fallacious.[12]

Evolutionary Moral Relativism

It is vital, Singer believes, that in discussing the matter of whether morality is one or many, we are clear about the level of morality we have in mind. If the pluralism that Wilson and I consider appears at the level of *rules of conduct*, Singer has no quarrel with us. But if it appears at 'the most fundamental level, the level of *ultimate moral principles*,' he must disagree.[13] Let us see how he fares.

As I mentioned earlier, Singer dismisses Wilson's first attempt to make good the claim that biology should take over ethics. Wilson was prompted to make this claim by his observation that moral intuition seems to be the final arbiter in moral philosophy. But moral intuition is ultimately based on the moral emotions that result from the operation of the hypothalamus and the limbic system. And that operation is itself the result of a long evolution. Neuroscience and evolutionary biology are thus two essential tools of the scientist turned moral philosopher. The first result of this approach is what Wilson calls moral pluralism, which stems from the realization that different human beings have different biological tendencies about altruism and other moral issues. Singer defends moral theory from such crude biological intrusions, and avoids the problems created by Wilson's biological relativism, by establishing his principle of impartiality. It is not a great problem for Singer that different humans have different tendencies or preferences on these questions. In resolving moral questions we simply have to take all such tendencies and preferences impartially into account.

Nevertheless, I suspect that Singer has not realized the strength of Wilson's position. We have already seen that Singer is forced to channel his principle of impartiality through the peculiarities of human nature. It is thus in principle possible that if human nature presents a fundamental pluralism, the divergence may be too great for one human morality to exist, even as an ideal. Wilson's moral pluralism would then be unavoidable. Let me clarify this point by discussing Rawls' Veil of Ignorance.[14] According to Rawls, a rational agent in the original position, looking through the veil of ignorance (that is, unaware of his social position and other personal characteristics) would choose a course of action, or an agreement that is fair to all sides, since an unfair decision may turn out to affect him most adversely. Under those conditions, rational agents (the requirement of rationality demands mainly that the agent choose consistently with his interests) would want, among other things, the maximum freedom consistent with equal freedom for others. They would also reject the principle of utility, for this principle requires that individual happiness should be sacrificed so as to produce the greatest balance of happiness over unhappiness — and on occasion that sacrifice may be extreme. Anyone who would agree to such an arrangement would be acting against his interest because he may be the one called upon to carry the unfair burden. In the opinion of Rawls, the principle of utility would cut too much against the grain of human nature.

If this account is correct, it seems that the knowledge recommended by Wilson is quite relevant to ethics. A radically different biology would much affect the outcome of the deliberations in the original position under the veil of ignorance. If, for example, rational ants existed, they would exhibit a group mentality; and thus would also, as Wilson imagines, 'regard individual freedom as intrinsically evil' (*EC*, p. 60). It is conceivable that such ants, writing their own *Theory of Justice,* would not come by Rawls' first principle of justice (of equal liberty) and that they may look favorably upon the principle of utility. Likewise, as we saw above, Singer's principle of impartiality may have to work through entirely different channels. As for human morality, the effect of these considerations depends on how divergent the biologies of, say, men and women happen to be. It may turn out that such divergence is not significant, but the point is that the issue is an empirical one.[15] From this result it seems to follow that moral philosophy must make empirical assumptions, and therefore that progress in moral philosophy requires certain appropriate empirical investigations. It appears, then, that biology is relevant to ethics at all levels. Singer will have to exert himself more if he is to cut Wilson off at the pass.[16]

In any event, as the case of the rational ants illustrates, evolutionary relativism appears to exist in ethics not merely at the level of *rules of conduct*, which is a significant result already, but at the fundamental level, the level of *ultimate moral principles*.

Towards a Future (Naturalized) Ethics

I would like to close with a sketch of the direction I favor for naturalized ethics. One issue I would like to explore is the use of the notion of relative truth advanced in Part I. In particular, I think that the application made in Chapter 4 to culture can be extrapolated to morality. In the case of the black murderer I spoke of a posture of hostility towards the world in trying to explain his pattern of conduct, a pattern that depended on several connections to his view of the world. I believe that this sort of causal approach can make better sense of our moral life than the logicist approach that employs moral values and rules as premises in moral situations reconstructed as arguments. The road to understanding does not lie along the trodden paths of logical inference, but in the exploration of the ways by which patterns of conduct that lead to a more fruitful social life come to be assimilated and of the means by which such patterns come to be recognized as appropriate in a given situation. Such recognition is a skill, or a set of skills rather, that permits, to differing degrees, the grasp of a moral situation, not only for action but for criticism (including attempts to justify our actions to others). The success of the performance that results, both at the level of moral perception and at the level of moral theorizing, is to be evaluated with respect to a social context, and thus the truth that may be 'revealed' to us will also depend on that context. Hence the possibility of applying to morality the notion of relative truth in a manner analogous to that of Chapter 4.

The fruitfulness of this proposal will ultimately depend on the neurological description of the brain in its moral 'mode'. It is thus encouraging to notice that recent developments in understanding the brain do suggest a compatible picture. In his wonderful book *The Engine of Reason, the Seat of the Soul*, Paul Churchland uses the latest developments of brain science to arrive at the conclusion that the best alternative to rule-based accounts of our moral capacity 'is a hierarchy of learned prototypes embodied in the well-tuned configuration of a neural network's synaptic weights'.[17] For instance, whereas the inferential approach gets mired by the fluidity of moral concepts (and thus an endless parade of exceptions and counter-examples), Churchland's explanation of concepts in terms of similarity gradients in the brain accounts for the flexibility of cognitive and moral concepts. It tells us, for example, why we can identify a cat as such even 'in the face of outright violations of almost every one of [the] conditions' of whatever definition of 'catness' we choose, whether commonsensical or scientific.[18] According to Churchland, 'One's ability to recognize instances of cruelty, patience, meanness, and courage. . . far outstrips one's capacity for verbal definitions of those notions'.[19] Churchland concludes that 'A relentless commitment to a handful of explicit rules does not make one a morally successful or a morally insightful person. . . the moral person [is] one who has acquired a complex set of subtle but enviable skills: perceptual, cognitive, and behavioral'.[20]

A full examination, let alone defense, of Churchland's views deserves a long and detailed new essay. My purpose in this short sketch is merely to indicate the apparent convergence of my evolutionary approach with his neuroscientific account of our moral faculty. What results appears to be a contemporary version of Aristotle's view on morality. My guess is that it holds much promise.

Notes

1. Hume, D., from Section VIII of *An Inquiry Concerning Human Understanding*, reprinted in Enteman, W.F., ed., *The Problem of Free Will*, Scribner's Sons, 1967, p. 182.
2. Bradie, M., *The Secret Chain: Evolution and Ethics*, State University of New York Press, 1994, p. 165.

3. *Ibid.*, pp. 166-167.
4. Singer, P., *The Expanding Circle: Ethics and Sociobiology*, Farrar, Strauss and Giroux, 1981.
5. Particularly Wilson, E.O., *On Human Nature*, Harvard University Press, 1978. For an important philosophical ally see Ruse, M., *Taking Darwin Seriously: A Naturalistic Approach to Philosophy*, Basil Blackwell, 1986. See also their joint work, 'Moral Philosophy as Applied Science,' *Philosophy*, 61, pp. 173-192.
6. Darwin, C., from Chapter XXI of *The Descent of Man*, reprinted in Appleman, P., ed., *Darwin*, Norton, 1979, p. 200.
7. *Ibid.*
8. *Ibid.*
9. Alexander, R., *The Biology of Moral Systems*, Aldine De Gruyter, 1987.
10. Richards, R., *Darwin and the Emergence of Evolutionary Theories of Mind and Behavior*, University of Chicago Press, 1987, p. 609.
11. Singer, P., 'The Expanding Circle: A Reply to Munevar,' *Explorations in Knowledge,* Vol. IV, No. 1, 1987, p. 52.
12. Some may harbor the suspicion that the social Darwinists must have committed the naturalistic fallacy at an earlier stage: when coming up with their mistaken pyramid of progress. But I suspect not. We value some states of the world more highly than others. But whether our valuations are sensible or not, the mere act of valuation does not in itself incur in a fallacy of reasoning.
13. Singer, P., *op. cit.*, p. 53.
14. Rawls, J., *A Theory of Justice*, Harvard University Press, 1971.
15. It may well be, of course, that social influences swamp the biological tendencies in humans, so we all end up with the same morality. Culture, however, can hardly be considered a savior against relativism — unless culture is somehow strongly anchored, given its well-known inclination to change. But what could be a suitable anchor if not biology?
16. Singer does try. He argues that Rawls' claim that the principle of utility would cut too much against the grain of human nature somehow assumes that human beings would not benefit as much by putting it into practice as they would from a different principle. But this, he thinks, shows that Rawls' claim assumes the principle of utility. ('Reply to Munevar,' *Explorations in Knowledge*, Vol. IV, No. 1, 1987). As I pointed out in the previous chapter, though, this argument misses Rawls' point. Rawls' point is that the principle of utility is not a reasonable way to achieve impartiality.
17. Churchland, P. M., *The Engine of Reason, the Seat of the Soul*, The MIT Press, 1995, p. 144.
18. *Ibid.*, p. 145.
19. *Ibid.*
20. *Ibid.*, p. 149.

12 A Naturalistic Account of Free Will

Introduction

Concerning the problem of the freedom of the will, Moritz Schlick once wrote that ". . . it is really one of the scandals of philosophy that again and again so much paper and printer's ink is devoted to this matter, to say nothing of the expenditure of thought, which could have been applied to more important problems. . ."[1] It was particularly distressing to Schlick to have to take up this "pseudo-problem," which he thought to be a problem at all only because of a misunderstanding, as David Hume with exceptional clarity (presumably) had already demonstrated.

I must confess a great deal of sympathy with Schlick's impatience. Nevertheless, like Schlick, I find myself having to address it, in part because I find neither Hume's solution nor Schlick's reiteration of it satisfactory, and in part because it seems especially daunting to the evolutionary naturalism that I have advocated elsewhere.[2]

The problem is usually thought to be that we are not morally responsible if determinism is true, for if determinism is true we are not really free agents. The reason is that determinism implies that every action is causally necessitated, but if so, we can never act otherwise, and if we cannot act otherwise we are not free.

The problem seems particularly acute for naturalist points of view, since naturalism would treat persons as arrangements of matter, and thus it is ultimately consistent with some form of determinism. Naturalism would, then, imply that free will is an illusion and so is moral responsibility. I think, however, that naturalism provides the most likely approach to the solution of the problem. This paper will thus aim to demonstrate that naturalism is compatible with the freedom of the will.

I shall begin by commenting on some recent naturalistic views on the problem. Although they fall short of the mark, they will prove to be helpful and instructive. I shall continue by discussing why indeterminism cannot account for freedom of action. I will then criticize some widely accepted solutions to the problem. And in the final section I will sketch a naturalist account of how the self determines the will, which is, I believe, the key to the solution.

Naturalist Views of the Problem

An intriguing example of a naturalist view can be found in Francis Crick's *The Astonishing Hypothesis*, in which he announces not only his theory of free will but also the location in the brain of the organ of free will.[3] Two aspects of Crick's approach are important to my discussion. The first is his theory. Crick sets out to explain what philosophers would call the "phenomenology" of free will, that is, why it appears to us that we have freedom of action. Crick starts from the sensible assumption that some part of the brain "is concerned with making plans for future actions without necessarily carrying them out."[4] We may, of course, be conscious of such plans. Crick further assumes, correctly I believe, that the actual working out of these plans (the "computations") are normally not open to our consciousness. We are aware only of the "decisions" taken by the brain, of the plans themselves. And finally, we are aware of the decision to act on one of these plans (e.g., to move), "but not of the computations that went into the decision."[5] Thus even if the workings of the brain are completely deterministic, this feature of consciousness bars our "direct" access to them, and therefore we are aware only of decisions untangled by deterministic mechanisms. In other words, even if the deterministic mechanisms are there, we cannot be aware of them, and thus we have the experience of acting "freely." Moreover, some of these mechanisms may be deterministic but chaotic (this idea he attributes to Patricia Churchland), which would make their outcome seem unpredictable.

The role of consciousness is much reduced in Crick's account (relative to the extraordinary significance philosophers characteristically assign to it). A brain so described, Crick believes,

> can attempt to explain to itself why it made a certain choice (by using introspection). Sometimes it may reach the correct conclusion. At other times it will either not know or, more likely, will confabulate, because it has no conscious knowledge of the "reason" for the choice.[6]

This result is consistent with the work by Crick and others on the relationship between consciousness and other functions of the brain, and with the points made in Part II to the effect that conscious deliberation is not always necessary for rational decision — let alone, as in this case, for mere decision, whether rational or not.[7]

Crick's hypothesis seems to lead to the conclusion that free will is an illusion. Others have arrived at this conclusion before, but not many have offered as compelling an explanation of *how* that illusion comes about. It seems to me that Crick's approach has one cardinal virtue and one cardinal flaw. Its virtue, which is a naturalistic virtue, is that it places some empirical constraints on our thinking about free will. Crick assumes that whatever free will turns out to be, it is a faculty of the brain, and thus studying the brain may reveal some of its important features. As soon as we take this approach, we realize, for example, the possibility that free will, as any other faculty of the brain, may malfunction; we realize, that is, that some people may have a defective, or non-existent, "organ" of free will. And indeed there appear to be such people, that is, people who seem not just unable but actually uninterested in choosing courses of action. Moreover, this handicap can be attributed to a lesion to a part of the brain that seems involved in making plans (the anterior cingulate sulcus, a region near the top and towards the front of the brain, which receives "many inputs from the higher sensory regions and [is] near the higher levels of the motor system"). As we will see below, this sort of hypothesis can be helpful in unraveling some of the puzzles of free will.[8] Nevertheless, Crick's explanation is flawed in that it leaves the main philosophical problem untouched. If we are not free agents, how are we then responsible for our actions?

Naturalists have tried several approaches to get around this problem. I will discuss some important ones. A rather common approach among scientists (as opposed to professional philosophers) is to analyze determinism in practical terms. E.O. Wilson, for example, describes the many cognitive abilities of a honeybee: it has memory, knows the time of the day, learns the location and quality of several flower fields, and responds vigorously and "erratically" to physical challenges. The bee thus

> appears to be a free agent to the uninformed human observer, but again if we were to concentrate all we know about the physical properties of thimble-sized objects, the nervous systems of insects, the behavioral peculiarities of honeybees, and the personal history of this particular bee, and if the most advanced techniques were again brought to bear, we might predict the flight path of the bee with an accuracy that exceeds pure chance.[9]

The point is that to human observers using such techniques the future of the bee is determined to *some* extent, but "in her own 'mind' the bee, who is isolated permanently from such human knowledge, will always have free will."[10]

In the case of human beings, we are so much more complex than bees that

> only techniques beyond our present imagining could hope to achieve even the short-term prediction of the detailed behavior of an individual human being, and such an accomplishment might be beyond the capacity of any conceivable intelligence. There are hundreds of thousands of variables to consider, and minute degrees of imprecision in any of them might easily be magnified to alter the action of part or all of the mind.[11]

Wilson suggests also an analog of Heisenberg's uncertainty principle, in which observations of human behavior alter that behavior. I suspect that he incurs here in an analog the mistake of claiming that we could never discover the nature of life because in studying it we alter it. In any event, all these reasons lead Wilson to suppose that no nervous systems may gain enough knowledge of the mind to "know their own future, capture fate, and in this sense eliminate free will."[12] You and I are consequently free and responsible in the fundamental sense that the "detailed histories of individual human beings [cannot] be predicted . . . by the individuals affected or by other human beings."[13]

This is not to say, of course, that we cannot predict general tendencies in behavior, or in fact what a particular person is likely to do. Wilson's concern is the ability to predict beyond statistical regularities. The scientific aspect of Wilson's analysis has been strengthened in the subsequent twenty years of great success in the neurosciences. As Paul Churchland explains, the brain is a non-linear system, that is, a system

> in which, at least occasionally, even the tiniest of differences in its current state will quickly be magnified into very large differences in its subsequent state. Since we can never have *infinitely* accurate information about the current state of any physical system, let alone a system of the complexity of a living brain, we are doomed to be forever limited in what we can predict about such a system's unfolding behavior, even if there are, and even if we happen to know, the inviolable laws that govern the system's behavior.[14]

Wilson's practical analysis, then, equates determinism with predictability and shows how perfect predictability is practically impossible. The first part of his analysis is by no means unreasonable. One of the oldest forms of the Problems of Free Will was precisely that of predestination: If God already knows how we are going to choose, in which sense can we be said to be free? A similar metaphysics of free will would be consistent with the views of those philosophers who take to heart Einstein's notion that time is an illusion. In the theory of General Relativity all events are described by four coordinates, time being one of them. Time becomes an objective coordinate whether we phenomenologically describe it as past, present or future. We might thus say that time is already laid out: Nothing can be except what is. For all we care, thus, our future actions are as objective as our past actions. We cannot deviate from what we are bound to do any more than are able to change what we have already done. I suspect that this view of time is done in by quantum physics (as is determinism itself, see below), but be that as it may, it is clear that some connection seems to exist between determinism and predictability. Wittgenstein, for example, claimed that freedom implies ignorance of what we are going to do (prior to deliberation, that is).[15]

Nevertheless determinism and predictability are not the same thing. As Churchland points out in the continuation of the passage quoted above, "Such systems are *strictly deterministic*, in the sense of being law governed, but they are nevertheless unpredictable, beyond their statistical regularities, by any cognitive system within the same physical universe."[16](My emphasis.) The problem of the freedom of the will involves determinism, not prediction. The reason is that, as William James saw, freedom requires alternative possibilities, but for determinism only what does happen is possible. But then, ". . . what sense can there be in condemning ourselves for taking the wrong way. . . unless the right way was open to us as well?"[17] It is for this reason that Wilson's suggestion ultimately fails. For in Wilson's practical freedom of the will our actions are still determined, whether we can predict them or not, and thus when we take the wrong way we do not really have the right way open to us.

There are those who would wash their hands off the whole problem and settle for some sort of practical "freedom." Churchland, for example, insists that the complexity of the brain (which creates the unpredictability) does provide us with the capacity for genuine spontaneous activity, for endless variety in our behavior, and this capacity is very important, even though

> It would be foolish to mistake such (genuine) unpredictability for what philosophers and theologians have often hoped for in the way of free will. That term was typically meant to apply to a human capacity that *transcended* the causal order, whereas the dynamical picture [of the brain] keeps us firmly embedded within the causal order.[18]

As sensible as these remarks sound, they would be philosophically more reassuring if we had an account of moral responsibility to go with them. Or if we could undermine the connection between determinism and moral amnesty. I will concentrate on the second of these options in the next two sections.

Indeterminism and Free Will

At first sight no discovery would seem to be as important for the issue at hand as the discovery that determinism is false. Thus we need to understand why the extraordinary success of quantum physics has not settled the matter. After all, quantum physics, at least in its orthodox interpretation, tells us that at the most fundamental level nature is to be seen as probabilistic, not because of ignorance (which is the traditional interpretation of probability in science) but because it is a basic property of physics. This is not to say that there are no deterministic processes or laws in the universe, but rather to deny that all phenomena are determined. This result raises the hope that the mind is not wholly determined, and that perhaps there is room for free will after all. As we will see in this section, however, this hope is misguided.

There are two main ways in which quantum indeterminism is introduced into the dispute. The first would have the (non-material) mind somehow interact with the brain at the quantum level. In this sense quantum physics would come to the aid of metaphysical (or substance) dualism, i.e., the view that mind and matter belong to separate "realities" or perhaps "dimensions." I presume this is what Churchland had in mind (no pun intended) when he spoke derisively of a human capacity that transcended the causal order. Dualism can take on three forms. First, the mind and its physical correlate (whether the brain or the whole central nervous system) exist in parallel dimensions: whatever happens in the mind has its counterpart in the brain. Second, the mind is an epiphenomenon of the brain, something like an innocuous halo on the causal order. Third, which was Descartes' favorite, the mind and the brain interact, viz., the mind interferes in the causal order.

Given the great recent success of the neurosciences in understanding the mind, the first two forms of dualism have little explanatory power and seem hopelessly *ad hoc*. In any event, neither epiphenomenalism nor parallelism can help with our problem of free will. Epiphenomenalism depends entirely on the causal (or not) order of the brain and thus neither adds nor subtracts anything to the issue of freedom. As for parallelism, if I raise my hand presumably of my own volition, but my action turns out to be determined, its counterpart in the mental dimension will also be determined. The perfect synchronization of the mental and physical dimensions will make it so that the circumstances that determine my brain processes (or else their counterparts in the mental realm) also determine my equivalent mental processes. This leaves us with the third option: interactionism, a view already saddled with having to overcome a great many implausibilities, chief among them the violation of the laws of nature (e.g., the conservation-of-energy law). The idea is, however, that quantum phenomena could come to the rescue here, for at the level of the extremely small perhaps the mind could guide the way these otherwise indeterminate events go. Once the mind has done its micro-job, the brain would amplify the quantum events and lead to action. Although I realize that the mind still is very mysterious, and that quantum physics also seems mysterious, I do not share the hope that they must thereby be connected. For in this connection we violate exactly the conditions that make quantum physics the paradigm of indeterminism. If mental events determine quantum events, if the latter are no longer fundamentally probabilistic, then quantum physics is false and we have not helped the cause of free will at all.[19]

These considerations show, incidentally, that we do not place ourselves in a better position to solve the problem of free will merely by eschewing naturalism and taking a higher, spiritual road instead.

A more naturalistic approach is to argue that brain processes have their origin in quantum events (which are amplified) and that therefore the brain is ultimately indeterministic. Unfortunately there are serious problems with this suggestion. One of them is that the acceptance of the suggestion would force us to accept also the notion that there are no deterministic processes whatsoever, for all physical processes can be said to have their origin in quantum events (e.g., when a billiard ball hits another, the electromagnetic fields of their orbital electrons first come into contact). Indeed, quantum theory draws a classical limit to which the statistical quantum properties tend, which means that outside of microphysics nature still is deterministic. Another problem, which I will not discuss here, is that the suggestions that have actually been put forward suffer from severe scientific implausibility as soon as we take into account not just physics but neurobiology (for a vivid example, see the devastating critique by Rick Grush and Patricia Churchland of Roger Penrose's attempt to link quantum physics and the brain via the microtubules of neurons).[20]

Ilya Prigogine has argued that complex systems, particularly those that are far from equilibrium, can be indeterministic. Even molecules in a real (non-idealized) gas respond to non-local and atemporal resonances, he claims, if nothing else because of their persistent collisions, which means that their motions depend on properties of the system.[21] This is not a plain emergent property of the system (i.e., that cannot be determined by knowing the properties of their constituents), it seems to me, but a stronger case in which the properties of the constituents are affected by the properties of the system. I am sure that Prigogine's interpretation of a system of gas molecules is open to dispute; what matters here, though, is whether the brain exhibits this strong property of emergence, and it seems to me that it does. Indeed, if anything the brain presents a clearer example, as we will see below. In systems far from equilibrium at least, Prigogine holds, we can in principle predict only statistical properties, and thus probabilities become a fundamental property of nature. I suspect that in developing this indeterminism born from instability and chaos, Prigogine has fallen into a subtle confusion between predictability and determinism, but this is not the place to do justice to his rather complicated analysis.[22]

Nevertheless, consider that the brain has about 100 trillion modifiable synaptic connections, that each of these connections can assume a large number of synaptic weights, and that these weights depend not only on the sensory signal arriving at the synapse of the neuron, say, and the structure of the neuron, but on the effects of other neurons on it. The sources of these effects, furthermore, are not restricted to the hundreds, or even thousands of neurons around it, but may be found also in neurons located in remote parts of the brain that connect in feedback loops to the neuron in question (indeed, as Paul Churchland points out, often what he calls recurrent pathways are more numerous than those carrying information forward from the senses). The result is that the brain systems or networks achieve temporary states of stability by "cycling through" many tentative synaptic weights and adjusting them so as to achieve a suitable accommodation to the goals of the moment. Thus a neural state is emergent in the sense that the weights of the synaptic connections that constitute it are not sufficient to determine it, and also emergent in the sense that those weights are also partially dependent on the neural state itself. But even though neural networks seem to fulfill Prigogine's criteria, every one of the relevant interactions obeys deterministic laws and, as the previous quotation from Churchland indicates, such networks should thus be considered to operate within the causal order. The brain is, then, both unpredictable and deterministic.

If perchance our actions were the result of indeterminism, of chance, the supporters of free will would face a most unwelcome discovery: Chance is also incompatible with free will. As Reid, and later Hume, pointed out, if our actions are the result of chance we have no control over them, they are thus not truly *ours*. In Gary Watson's words, "What destroys freedom . . . is the lack of self-determination and that results both when the will is determined by other events or states of affairs and when it is not determined at all."[23] Watson then puts his finger on the true nature of the problem: "The negative requirement that the will not be causally necessitated by antecedent events is dictated by the positive requirement that the will be determined by the self."[24]

Naturalists, scientists in particular, have been looking in the wrong place, their quest has been distorted by their emphasis on only one or two aspects of the problem: predictability, determinism, or both. Our task should instead be to provide a naturalistic account of how the self determines the will. But before this task is undertaken more directly, one more hurdle must be overcome: the claim made by some philosophers that there is really no problem of free will at all. Otherwise we would spend even more time and effort on a task for which there is no need.

Philosophical Solutions

Many philosophers of science tend to ignore the problem of free will because they think that Hume already solved it a long time ago. I disagree with them . Hume's first move was to undermine the notion of Necessity (the inevitability that James understood was part and parcel of determinism). Our notion of necessity and causation, says Hume, "arises entirely from the uniformity observable in the operations of nature, where similar objects are constantly conjoined together, and the mind is determined by custom to infer the one from the appearance of the other."[25] We are thus wrong when we believe we "perceive something like a necessary connection between the cause and the effect."[26] This is very important, for the regularity of human behavior is crucial to our social lives. Indeed, Hume asks, "Where would be the foundation of *morals*, if particular characters had no certain or determinate power to produce particular sentiments, and if these sentiments had no constant operation on actions?"[27] Thus we are wrong when we suppose that a difference exists between "the effects which result from material force" and those "which arise from thought and intelligence."[28] As Hume puts it, "this experimental inference and reasoning concerning the actions of others enters so much into human life that no man, while awake, is ever a moment without employing it."[29] But in both cases "we know nothing further of causation of any kind other than merely the *constant conjunction* of objects and the consequent *inference* of the mind from one to another."[30]

A solution cannot be found, Hume argued, "as long as we will rashly suppose that we have some farther idea of necessity and causation in the operations of external objects."[31] while in the case of human action "we feel no such [necessary] connexion."[32] Once we realize that the same notion of "necessity" operates in both, and that there is nothing really necessary about causation, the way is open to a solution. And the solution is simple: when we think carefully about cause and effect we come up with an innocuous connection that does not conflict with freedom of action. Such a conflict comes about because of our rash judgment about causation and because of one additional misunderstanding: our opposing liberty to necessity and not to constraint. Hume follows Hobbes in claiming that by liberty, when applied to voluntary actions, "we can only mean *a power of acting or not acting according to the determinations of the will*; that is, if we choose to remain at rest, we may; if we choose to move, we also may."[33] The question is settled, then, for "this hypothetical liberty is universally allowed to belong to every one who is not a prisoner and in chains."[34]

Hume's solution has been adopted by a variety of philosophers. Mill, for example, tells us that "Even if the reason repudiates, the imagination retains the feeling of some more intimate connexion, of some peculiar tie, or mysterious constraint exercised by the [cause] over the [effect]."[35] It is precisely in that mysterious constraint that the conflict arises, for "We know that we are not compelled, as by a magical spell, to obey any particular motive."[36] Schlick calls our attention to the distinction between natural law and the law of the state. The first one involves determinism and its "necessity" consists merely in its being universally valid. The second involves compulsion. The problem of free will stems from the confusion of the two, that is, from ascribing to natural law the compulsion characteristic of a law of the state. For him, as for Mill and Hume, and many more recent philosophers, the entire controversy can be avoided if we only pay proper attention to the meaning of words. And it is in that proper attention that we understand that causality, far from being incompatible with responsibility, is required by it. "We can speak of motives only in a causal context," Schlick says, "thus it becomes clear how very much the concept of responsibility rests upon that of causation, that is, upon the regularity of volitional decisions."[37] Hume's very point.

Nevertheless, this approach to the problem is found wanting when we remember William James' simple observation: we cannot be blamed for taking the wrong way when the right way was not open to us. Let us look first at the much-debated Humean account of causality. Suppose that a wrecking ball is suspended ten meters over a delicate porcelain vase. If we let go of the ball it will pulverize the vase. Any deviation from this result will require special circumstances (e.g., a hidden gigantic electromagnet). But if all the relevant circumstances are accounted for, and they are as first described, when the ball is set loose it *will* destroy the vase. It *will* happen that way, to the exclusion of alternatives. Hume himself would not deny this point. We thus need no belief in mysterious constraints or magical spells to realize that determinism leaves no options open. Second, we may realize that the problem does not arise from confusing natural law and the law of the state, that is, from attributing compulsion to natural law (or causality). On the contrary, compulsory law may give us an excuse, or a justification, for not acting otherwise, but we seldom, if ever, attribute to it a universal character. Sometimes we disobey it, and there are laws that most people disobey (e.g., speed laws for automobiles). The problem with acting in accordance with natural law (which we cannot avoid) is precisely that it places us in the same category as the wrecking ball, regardless of how special and free we feel. Einstein once wrote that

> If the moon, in the act of completing its eternal path round the earth, were gifted with self-consciousness, it would feel thoroughly convinced that it would travel its path on its own, in accordance with a resolution taken once and for all.[38]

Compulsion may exculpate us, but it allows us to remain moral agents. If a man with a gun to his head is told to kill another, and complies, we excuse him. But if he is thrown from a tall tower and thus kills another on the ground, the question of excusing him does not come up, for being a falling body is completely out of the moral sphere. Therefore Hume's, Mill's, and Schlick's "proper attention" to the meaning of words does nothing to solve the problem of free will. The problem is created by determinism precisely because it seems to put our actions on a par with the behavior of falling bodies — precisely because when we act "wrongly" we do not seem to have the "right way" open to us.

Nor is further "linguistic" analysis of the "logic" of freedom and responsibility likely to improve matters much. The reason is simply that such analysis belongs ultimately to the "phenomenology" of the issue, even if limited to the level of discourse, and appeals to such phenomenology are rendered moot at best, and question-begging at worst, in a dispute in which the challenge is to show that it is not an illusion. Imagine an extremely sophisticated android who, unbeknownst to him, receives radio instructions from me as to how to behave (a suitably wired human would do also). In accordance with Crick's account of the brain, the android is aware only of the "decisions" (stand, sit, etc.) and not of the fine work of his brain's mechanisms (Crick's "calculations"). Unlike my brain, however, his brain is affected by my radio signals and the decisions it arrives at are chosen by me. If I instruct him to remain at rest, he may; if I instruct him to move, he may. That is, nothing prevents him from carrying out the action that his consciousness informs him is his brain's decision — his will — for he "is not a prisoner and in chains," as Hume said. Or to use Hobbes' notions, once he "wills" something he has the power to do it. The android is then a free agent because "he can do if he will, and forbear if he will."[39] The android may thus satisfy what Hobbes, Hume, Mill, and Schlick thought we "mean" by freedom of the will, but he is not really a free agent because his will is determined by me.

Other conceptual and linguistic analysis may perhaps uncover different important "meanings" concerning freedom and responsibility, but we will not make much headway with our problem as long as the much despised "free-will metaphysician" can claim that our will is determined by something other than its own self.

This point extends to another "discovery" of linguistic philosophy, namely the claim that there is no problem of free will because the language of determinism belongs properly to material things, whereas the language of thought, which presumably covers such notions as freedom and responsibility, involves not causes but reasons and intentions. But as Gary Watson argues, the traditional problem is not avoided in this way.

> For if physical determinism is true, it is impossible for your body to move in any way other than its actual motion. This means that it is not possible that you will *move* your body in any different way, and hence that you will act in any way that requires a different bodily motion.[40]

Moving the discussion to the realm of reasons and intentions still does not explain how we can act otherwise. As Watson continues, "If it is physically determined that your arm does not go up during a certain period, then it is not possible that you will signal the waiter, say, by raising your arm." Unless, of course, this conceptual dualism, as we may call the philosophical view under discussion, manages to explain "how it can be the case that you are able to raise your arm during a time when it is causally impossible for your arm to go up."[41]

Naturalism and Self-Determination

The question is whether the will can be determined by the self. I believe that it can, that it is typically so determined, and that we can understand how once we adopt a naturalistic approach. In the rest of this section I will sketch such an approach, with enough detail, I hope, to constitute a good first approximation.

Let us begin with a naturalistic account of the self. Most, if not all, living things have at the very least a primitive sense of self. A single-cell organism, for example, is organized to distinguish what belongs to it from what does not. Indeed, most bacteria will identify invading cells and use a variety of chemical means to destroy those invaders. In a primitive sense we may say that organisms can tell self from non-self. As the complexity of the organism increases, a nervous system may develop to coordinate the different organs and functions. However loose this organization may be in some "primitive" animals, it is more than a mere black box by means of which the inputs to sensory receptors are transformed into behavioral outputs. The behavioral response, if any, is modulated by the animal's sense of itself and of its relation to the world, a rather complex achievement that depends on internal information (gathered through proprioceptors) and its mutual fine-tuning with the external sensory information. As the complexity of the nervous system increases, so does the number of possible ways of coordinating and modulating neural systems. Visual perception, for example, depends on feedback loops from the other senses, including the internal senses, as well as from modalities that involve memory, expectation and even emotion. This increase in complexity leads to an increase in plasticity and to the appearance of what we might call intelligence. In social animals presumably the brain is genetically "wired" if not for cooperation, for the capacity to cooperate, as Darwin argued over a century ago.[42] This disposition toward cooperation, made up of our "social instincts," leads to the origin of morality in intelligent creatures. In a complex brain, of course, the moral emotions that may result from such a disposition are subject to a great amount of plasticity. It is from those emotions, in any event, that, according to Darwin, we derive our moral conscience. In this connection, with the development of moral responsibility in the brain, we should find the relevant features of free will. After all, the problem of free will revolves around the question of whether the decisions we take concerning those actions thought to fall within the sphere of moral responsibility are determined by the self or not.

In this naturalistic account the self is, of course, none other than the brain. So the question is whether the brain determines the will, i.e., whether it has the power to make those decisions open to moral evaluation. The answer is that the brain does. Lest we think that this answer is too facile, let us realize that the brain's self-organization pits the organism as an independent entity vis-a-vis the world. When discussing Prigogine's claims against determinism, which he based partly on his analysis of the order exhibited by many states far from equilibrium, I brought up the notion of strong emergence and explained how it applied to the brain. This strong emergence goes beyond complexity and plasticity in establishing the operation of any particular brain as *sui generis*, since any small variation from brain to brain — and there are large variations — can be greatly amplified, as Churchland pointed out above, and since a brain will work through a given situation in *its own* mysterious ways. The question here is not practical, or even in-principle, unpredictability, as it was for Prigogine, but selfdetermination.

Although the world exerts an influence on the brain's decisions, either as present stimuli, or as past experience of the species (which gives us our basic modes of thought and perhaps our basic moral emotions), a strongly emergent system such as a human brain amounts to a pocket of the world ruled by emergent "laws" of its own. There may be a worry about whether we may speak of universal laws in the case of the brain with any greater assurance that we can in the case of evolutionary biology, but there is no question that the brain exhibits emergent causal relationships and that its causal system is generally self-sustaining and independent. Again, by independent I mean simply that there is a discontinuity between the rest of the world, its natural laws included, and the new emergent "laws" by which each individual brain interprets a situation, finds it relevant, evaluates it, and finally decides how to deal with it. Natural laws, of course, operate in the brain, with the many elements that come to play a role in any decision organized in a manner that roughly resembles that of other brains, but that also depends on the peculiar characteristics of each brain. It is that organization that places the whole of those elements beyond the behavior of mere falling bodies, just as it is the organization of the elements of the brain that makes their joint action intelligent. That is how intelligence arises out of matter: as a peculiar systemic property of a peculiar dynamic organization of that matter. That is indeed how mind, the self, arises out of matter. We may have this image of the little cogs and wheels of the brain blindly turning inside our brains in obedience to indifferent and universal laws, but this image ignores that these very cogs and wheels are not external factors impinging on the self but internal elements of a complex whole which is , to borrow the cliche, more than the sum of its parts. Moreover, not only are these cogs and wheels elements of the self, but their very character as elements is determined by the complex whole to which they belong. Neurons, for example, have their synaptic weights modulated by a complex array of influences from other neurons, as Paul Churchland has explained in his account of recurrent pathways in the brain.[43] It is reasonable to suppose that complex arrays of information higher in the brain's hierarchical structure (what I take to be Wilson's "schemata") interact, combine, and compete with other such complexes for central stage. The contribution of an element of the self to a decision then depends partly on the systemic influences of the self on that element.

How is this a solution to the problem of free will? To borrow Watson's words once again, "the question is how a series of natural processes (for which you are not accountable) can result in processes and events over which you do have control (for which you are accountable)."[44] Let us, then, see what sorts of processes and events we do no have control over and how they result in some over which we do control (and for which we are thus accountable). Schlick fairly ascribed to the free-will metaphysician the view that the will is determined by character and motives, and therefore we can do nothing about the way our decisions go, for we have no power over either: "the motives come from without, and my character is the necessary product of the innate tendencies and the external influences which have been effective during my lifetime."[45] Many motives, however, are internally generated — think of ambition or self-loathing, for example. And those which we may regard as external do not come entirely from without either. A hostile look from a stranger motivates me to search quickly for a heavy object that I can hold in my hand. But it does so only because I interpret his look as hostile, and because I read the situation as dangerous since there are no other people around, my foot injury will keep me from running, and so on. Something does not become a motive unless it is first interpreted by the brain in a particular way and then seen as relevant to a goal that the brain has decided to pursue. External influences do exist, of course. Almost everyone in the world would consider a pistol pointed at his head a motive for action, but even then the pistol has to be recognized as such (not a toy pistol, for example), other clues have to be read as corroborating the impression of danger in the situation, etc. However we interpret the situation, though, we may have no control at all over the event of having the pistol pointed in our direction. Once the event is processed and integrated into our mind it provides a motive and we may make a decision about it. But this decision is guided by the characteristics of the particular brain, by its needs, desires, beliefs about past experiences (which would include true and false memories), by its emotions, and by the urgency of other decisions taken about the same time. Every decision is thus stamped with the very personal seal of the brain that has control over it. Some people would run, others would shoot first if armed, and still others would take the bullet to protect a loved one.

As for character, the pertinent external influences would be handled in the same way just described. The innate tendencies can indeed be strong: evolution and embryology are certain to forge in the brain a variety of dispositions, but the brain is also distinguished by its ability to adapt, to change, to learn, and thus to transform itself. Moreover, this transformation is to some degree a matter of choice. In this, character is not an exception. As Mill pointed out, a man's "character is formed by his circumstances ... but his own desire to mould it in a particular way is one of those circumstances, and by no means the least influential."[46] Indeed, Mill suggests that the feeling "of our being able to modify our own character *if we wish* is itself the feeling of moral freedom which we are conscious of." It seems to me that Mill is right at least to this extent: we do blame ourselves, at least in moments of candor, for bad character traits that we realize we could have overcome through firmer resolve as we formed our habits. Many internal processes may exert great influence not only on the formation of character, but on the decision making itself. But we should keep in mind that those processes, too, do not become factors in a decision until they are assimilated into the whole by emergent mechanisms. When the self determines a decision, it does so *qua* self, its choices are not forced upon it by factors over which it has no control. On the contrary, the self *qua* self is what controls all these factors, assigns them values within the system, makes them relevant, compares and combines them with other factors. Otherwise they would play no role in the decisions the self makes.

It is precisely when factors outside this organic assimilation and control by the brain determine a decision that we can correctly claim lack of moral responsibility. Let us apply the previous android test. If Peter (Peter's self) would have decided to stand up, but through radio signals I alter the decision so as to stay seated instead, Peter is not acting freely. It was not his brain that made the choice, even though the brain mechanisms may be such that he cannot help but think that he made the decision himself. Likewise with drugs, brain injury, or any disease that alters the normal decision-making operations of the brain. When a disruption in the proper rates of neuro-transmitters render Mary completely unable to interpret a situation as she would under normal circumstances, or gives extraordinary significance to an event that would not be normally that important to her (as in drug-induced paranoia), we should exempt (not merely excuse) her from moral blame to the degree of her inability. Of course, we may blame her for the choice to take the drug, and therefore charge her with negligence, but the reason for this harsh judgment is that her brain was working normally then. And when Alzheimer's robs John from access to his past, when the continuity of his self is thus disrupted, we are again not entitled to assign blame.

When a man is himself, his decisions will not proceed from character and external influences with necessity. They will result instead from the integration of internal and external influences into an individual-specific, and strongly emergent causal system: his self. Naturalism thus allows us to conclude with confidence that the will is determined by the self.

Unlike Hume's, Mill's, and Schlick's notions of freedom, all of which fail the android, drugs, injury, and disease tests, this naturalist account of the will (the self's — the brain's — power to make decisions) does a better job of explaining the phenomenology of freedom and responsibility. It does so while explaining how events and processes over which we have no control are assimilated by the self (the brain) and as a result we engender processes over which *we* do have control and for which we are, therefore, morally responsible.

Notes

1. Schlick, M., "When is a Man Responsible?" reprinted in Enteman, W.F., *The Problem of Free Will*, Scribner's Sons, 1967, p. 184.
2. See for example "Evolution and Justification," *The Monist*, Vol. 71, No. 3, 1988, pp. 339-357, and *Radical Knowledge*, Hackett, 1981.
3. Crick, F., The Astonishing Hypothesis: The Scientific Search for the Soul, Scribner's Sons, 1994, pp. 265-268.
4. *Ibid.*, p. 266.
5. *Ibid.*
6. *Ibid.*
7. Until rather recently scientists tended to place the same importance on reflection. As late as 1973, Salvador Luria wrote that "Human behavior is conscious behavior and by virtue of that fact man is more than another animal." (*Life: The Unfinished Experiment*, Scribner's Sons, 1973, p. 146.) Darwin himself thought

that the moral sense comes from the application of the higher mental powers to the social instincts, and by those powers he meant memory, anticipation, and the power of reflection (cf. Bradie, M., Note 41.). It is true, of course, that deliberation often amounts to conscious deliberation, but it need not be so. Think of the hundreds of little, and sometimes big, decisions that we make in the course of the day: while driving, walking, taking the stairs, dancing, painting, or playing a game. Some may be the results of habit, automatic "subroutines" of the brain, but in many cases we have to own up to them. In sport, for example, good decisions often depend on our reading the situation correctly and quickly and making a "split-second" choice appropriate to the situation. In war those quick decisions could not have more serious consequences. It would be strange to disown those decisions, to say that we did not want to act that way, or that we are not responsible for them but will accept responsibility only for those actions about which we deliberated at length.

8. A result of this line of thought is that the question of free will may extend into the animal kingdom well beyond humans (cf. Wilson's discussion of honeybees below), for it may be difficult to draw a sharp line of demarcation, as is the case with other traits that have resulted from natural selection, including intelligence (which does not require, by the way, that mice, say, ask themselves whether they are intelligent). A similar approach would best fit the emergence of consciousness. Some philosophers may think that language is necessary for conscious reflection, but Paul Churchland has made an excellent case for the claim that "[T]he cognitive priority of the preverbal over the verbal shows itself, upon examination, to be a feature of almost all of our cognitive categories." (*The Engine of Reason, the Seat of the Soul*, MIT Press, 1995, p. 144.) An advantage of naturalism is the possibility of future comparative animal studies that would permit us to understand better human decision making by seeing how it differs from animal decision making.

9. Wilson, E.O., *On Human Nature*, Harvard University Press, 1978, pp. 72-73.
10. *Ibid.*, p. 73.
11. *Ibid.*
12. *Ibid.* p. 74.
13. *Ibid.*, p. 77.
14. Churchland, P.M., *op. cit.*, p. 113.
15. See the discussion by Paul Horwich in his *Asymmetries in Time*, MIT Press, 1987, p. 204.
16. Churchland, P.M., *op. cit.*
17. James, W., "The Dilemma of Determinism," reprinted in Enteman, *op. cit.*, p. 69.
18. Churchland, P.M., *op. cit.*
19. Another possibility would be to have the non-physical mind somehow collapse the wave packet. I have always found difficult to understand suggestions to that effect.
20. Grush, R., and Churchland, P.S., "Gaps in Penrose's Toilings," *Journal of Consciousness Studies*, 2, No. 1, 1995, pp. 10-29.
21. Prigogine, I., *The End of Certainty: Time, Chaos, and the New Laws of Nature*, Free Press, 1997.
22. Prigogine, following Poincare, argues that the atemporal resonances make it impossible to integrate over the trajectories of the gas molecules. Presuming that

this is right, we still need to consider whether the non-integrability amounts to in-principle unpredictability. But even presuming that it does, from non-integrability we cannot conclude indeterminism, nor can we do so from unpredictability alone, as we have seen.

23. Watson, G., "Free Will," in Sosa, E., and Kim, J., eds., *A Companion to Metaphysics*, Basil Blackwell, 1994, p. 178.
24. *Ibid.*
25. Hume, D., "Of Liberty and Necessity," from *An Enquiry Concerning Human Understanding*, Open Court Publishing Co., 1907. Passages quoted in this essay come from Section VIII, Parts I and II, reprinted in Enteman, W.F., *op. cit.* This quote comes from p. 166.
26. *Ibid.*, p. 174.
27. *Ibid.*, p. 172.
28. *Ibid.*, p. 174
29. *Ibid.*, p. 172.
30. *Ibid.*, p. 174.
31. *Ibid.*, p. 175.
32. *Ibid.*, p. 174.
33. *Ibid.*, p. 176.
34. *Ibid.*
35. Mill, J.S., "Of Liberty and Necessity," from *A System of Logic Ratiocinative and Inductive*, Longmans, Green, Reader and Dyer, 1872, II, reprinted in Enteman, W.F., *op. cit.*, p. 257.
36. *Ibid.*
37. Schlick, M., *op. cit.*, p. 192.
38. Quoted in Prigogine, I., *op. cit.*, p. 13.
39. Quoted in Watson, G., *op. cit.*, p. 176.
40. *Ibid.*, p. 180.
41. *Ibid.*
42. See the discussion by Michael Bradie in his *The Secret Chain: Evolution and Ethics*, SUNY Press, 1994, pp. 58-64.
43. Churchland, P.M., *op. cit.*, pp. 97-150.
44. Watson, G., *op. cit.*, p. 181.
45. Schlick, M., *op. cit.*, p. 186.
46. Mill, J.S., *op. it.*, p. 260.

PART IV
APPLICATION TO SPACE SCIENCE

13 A Philosopher Looks at Space Exploration

Introduction

The purpose of this paper is to offer a philosophical justification of space exploration. There are two reasons for undertaking this task. The first is that little has been done on the way of a philosophical examination of the field (by professional philosophers at any rate), even though one of philosophy's main concerns is the justification of all important human enterprises. The second is that the standard justifications offered for space exploration are not truly satisfactory, and that philosophy, particularly philosophy of science, can be very helpful in this regard.

Space supporters are often perplexed by the failure of their fellow citizens to grasp the significance and fascination of exploring the heavens. It seems to me, looking in from the outside, that greater success can come only from paying greater attention to the reasons the other side may have for resisting. True argumentation cannot take place in a vacuum. You may not always be able to convince even open-minded people, but you improve your chances when you consider seriously what keeps those people from coming over to your side. This is why impatience tends to be rhetorically self-defeating. In the words of Arthur C. Clarke (1946): "the urge to explore, to discover, to follow knowledge like a sinking star" is its own justification. This self-justification is presumably rooted in human nature or human destiny. As the Norwegian explorer, Fridtjof Nansen once said, "The history of the human race is a continuous struggle from darkness to light. It is therefore of no purpose to discuss the use of knowledge — man wants to know and when he ceases to do so he is no longer man." (As quoted in Greve *et al, 1976.)*

This attitude may be all well and good. But why should anyone who is not already convinced accept it? Even if our fellow citizens thought that we are naturally inclined to explore, they may remain unmoved by our aspirations. For the fact that an inclination is natural does not make it good. On the contrary, to become morally good we often have to learn to curb some of our natural inclinations. Most philosophers would even apply the term naturalistic fallacy to what seems like an attempt to derive values from facts. I myself think that there is something right about the intuition that, at least where it comes to knowledge, natural inclinations must be taken into account (Chs. 9 and 10), but that topic is beyond the scope of this paper. Suffice it to say for now, that space supporters who eschew justification are unlikely to convince the rest of the citizenry.

With this considerations in mind, I will first describe the most important objections to space exploration and the standard replies to them. As we will see, this standard case for exploration does not quite meet the challenge. It is precisely at that juncture that philosophy will prove most useful.

The Objections

There are basically two main kinds of objections that must be overcome to justify space exploration: social and ideological. The social critics argue that space exploration takes money, talent, and effort away from more pressing human needs (such as combating poverty and hunger). The ideological critics hold that space exploration is an unwise activity, an extension of the big science and technology that, coupled with the mentality of growth, have done so much to destroy our environment, deplete our resources, and bring our planet to the brink of disaster. The problem is presumably with the scientific mentality that leads us to exploit and interfere with nature, instead of trying to live in harmony with it.

Reply to the Objections: The Standard Case

In reply to these objections, space enthusiasts tend to describe the many benefits we derive from the space program: weather satellites save lives and crops, communication satellites bring about an economic expansion, land satellites discover resources and help us monitor the environment, and space technology spins off valuable products into our lives. Space exploration today contributes greatly to the reduction of human misery, the improvement of human life, and the preservation of the environment.

Moreover, we are just beginning to move into space. The low gravitation and the vacuum of space offer many industrial and technological advantages. And in the solar system we may find new wealth and new sources of energy for our planet (e.g., collectors of solar energy — solar power satellites — that would provide electricity for the Earth), thus helping us solve our industrial and environmental problems at the same time.

In summary: space exploration already goes a long way to meet the critics concerns. If given a chance, space will make possible a richer, cleaner and more humane future. This is the basic standard case in favor of space exploration.

This basic case used to be buttressed with economic good news: econometric reports allegedly showed that space returned to the economy far more than we invested in it. One common figure was that for every dollar we spent on space we got seven back. Perhaps the reports reflected the actual economic performance, but unfortunately the econometric studies suffered from a variety of shortcomings (e.g., they ignored that knowledge is the main result of exploration, that many of the inventions in question were influenced by many fields, and that social factors are very important in economic growth; furthermore, generally the formulas used were from previous studies on military research and development (R&D) in the fifties, with the space expenditures plugged in, not actual space R&D economic performance). The economic case was thus suggestive — we do see many industries and technological products that owe their existence to space — but not conclusive. (Holman, 1974) That was in the Golden Age of exploration (1962-1976), in any event. Today the Space Shuttle makes it so expensive to place things in orbit that the economic case is not quite as strong as it once seemed. It has been said that if the alchemist dream of turning lead into gold could be realized simply by taking the lead aboard the Shuttle, it would cost more than just buying the gold! Eventually we might be able to bring the costs down significantly, but given the promises made about the Shuttle and the actual disappointing performance, we cannot expect the general public, let alone the critics, to be persuaded.

Nevertheless, in spite of this concerns about its economic aspects, it does seem that the standard case meets the objections quite successfully. Why do critics still persist then? To see why let us see the case in outline form:

SATELLITES:
    WEATHER:
        save lives
        help agriculture
        help transportation
    SEA:
        find resources
        tell us about environmental impact
    LAND:
        find resources
        tell us about environmental impact
    COMMUNICATIONS:
        help commerce
        make our lives easier

SPINOFFS:
NEW TECHNOLOGIES
NEW TECHNOLOGICAL OPPORTUNITIES

The Counter Reply

Ideological critics. When space supporters claim that space exploration can help solve our environmental problems, the ideological critics are not impressed. For them, space is a delusion, for it offers more growth and technology to stop the mess caused by growth and technology. Of course, the more we foul up the world, the more space will look like a necessity. But this is a false technological panacea. It is rather like a pain reliever that keeps the patient from having the operation that will save his life. As Wilson Clark (1976) puts it when criticizing Gerard O'Neill's advocacy of space colonies and solar power satellites, "[he] speaks in terms of a 'first beachhead in space,' evoking the image of greener grass on yonder hill. Unfortunately, we have little time in which to prevent the elimination of the vegetation altogether."

According to the ideological critics what we need to do is to change our attitude and stop fouling the environment. If we do that, we won't need space technology to help solve the environmental crisis.

Social critics. Many important space activities do not have the obvious beneficial consequences of weather and communication satellites. Where is the obvious payoff of a probe of Jupiter or Titan, of landing a vehicle on Mars, of scooping a bit of Halley's comet? Few accomplishments of space exploration rank as high as the discoveries made with telescopes in orbit. But how is that information from space astronomy going to put food in children's mouths or a roof over their heads?

In emphasizing the practicality of space technology, the standard case makes an orphan of the heart of space exploration: those exciting activities motivated by our sense of adventure, by our urge to explore, by the need to satisfy our curiosity. That is why the standard case's attempt to justify exploration along practical lines does not go far enough.

Moreover as far as many social critics are concerned, there is another serious objection: if spinoffs are so valuable, does it not make more sense to spend the money directly in the relevant fields?

The Serendipity of Scientific Exploration

In trying to overcome this new round of objections, space enthusiasts point out that when we first explored space we did not know for certain that so many good things would repay our efforts; very often we had no inkling.

The pursuit of scientific exploration pays because of the *serendipity* of science; that is, because of the unintended benefits that science yields.

Unfortunately the critics may doubt that the prior performance of the space program is enough to guarantee serendipity. Having gotten water out of a well before does not guarantee an inexhaustible supply.

Enthusiasts point to examples of serendipity in the history of science, but they need to offer more than the customary list of confirming anecdotes. For example, it is not enough to remark that the research on electromagnetism by the 19th Century Scottish physicist Clerk Maxwell made possible television and computers, two inventions which Maxwell himself could not have foreseen. Unfortunately this is very one-sided historical analysis, for little is ever said about the overwhelming majority of the research carried out during the 19th Century. Did all of that science yield practical benefits, or only some of it? Some of space exploration has had practical consequences, but not all of it. And it is difficult to see how the exploration of Jupiter will help us here on Earth. How can we guarantee the serendipity of our future space science?

A Philosophical Case for the Serendipity of Science

Neither history, nor economics, nor the natural sciences seem to provide us with a justification that is rhetorically satisfactory, but I believe that philosophy of science can. One of the main purposes of philosophy of science is to analyze the nature and limits of science, and in this particular regard the issue is whether there is something about the nature of science that makes serendipity practically unavoidable. For if serendipity is a natural consequence of science, then science will be practical in a very profound way, and we will have an answer to the concerns of the social critics. As it will turn out, this philosophical examination of the nature of science will also allow us to meet the main ideological objection.

The argument has two parts. The **first part** establishes a strong connection between scientific change and serendipity. It goes as follows.

1. Scientific views are instruments for interacting with the universe. According to a still popular view, scientific knowledge is objective, objective is equated with factual, and as a result facts become the business of science. From this it presumably follows that the function of science is to collect facts about the universe, and that the function of space science is to collect them from space.

Developments in the philosophy of science during the second part of the century offer a much more complex picture of science. One lesson we learn from contemporary thinkers like Paul Feyerabend (1975), Thomas

Kuhn (1970), Imre Lakatos (1978), and to some degree Karl Popper (1959), is that scientific views or theories are like spectacles by which we experience the universe. We might even say that they are instruments through which we conceive of the universe. According to Kuhn, for example, scientific views (which he calls "paradigms") tell us what elements there are in the world, what relations exist between those elements, and thus what sorts of problems are meaningful in a particular science, as well as what kinds of answers are acceptable (for a scientific view determines the theoretical, mathematical, and experimental commitments of the discipline). To speak of science as a pair of spectacles that permit us to see the world is to speak metaphorically, to be sure. But that metaphor is by no means farfetched. We should realize that without our scientific views we would simply be blind to many aspects of the universe. And those views give us more than pictures of the universe: they also provide means of interacting with it.

This means that science gives us more than mere representations of the universe. For science asks questions by seeing, hearing, analyzing, probing, and touching nature at many different energies and magnitudes. Thus when we learn to "see" the universe we actually learn to make contact and deal with its diverse facets in many different ways. To illustrate how far the shift on emphasis away from the collection of facts can take us, let us consider briefly the manner in which Galileo surmounted one of the main obstacles to Copernicus' idea of the motion of the Earth.

In his analysis of the history of science, Paul Feyerabend, explains how the Tower Argument was employed to show that the "facts" refuted the Copernican view (see Appendix B). Suppose with Copernicus that the Earth moves. If you then drop a stone from the top of a tall tower, by the time the stone hits the ground the tower will have moved with the Earth, and thus the point of impact will be a considerable distance from the base of the tower. For the impact to be as close to the base of the tower as it actually is, the stone should follow an oblique motion. But it obviously falls straight down. Thus the supposition that the Earth moves cannot be correct.

We now know, however, that Copernicus was right: the Earth does move. But why should his view fly on the face of such obvious facts as the straight downward motion of the stone? As we see the stone leave the tower, we find it natural to say that the stone moves straight down. But this "natural interpretation" of what we see assumes that the motion of the stone can be determined by a normal observer under normal conditions. "Real" motion is observable motion, a change in location that we can measure (in this case a normal observer functions as the measuring instrument: the motion of the stone is what he sees the stone do). In terms fashionable today, the opponents of Copernicus and Galileo assumed that motion was operational (determinable by measurement) and absolute.

Galileo's ploy was to offer a different set of "natural interpretations" according to which we do not observe the real motion of the stone. As it turns out, the stone, the tower, and the observer share circular inertia with the earth. There are, then, two components to the motion of the stone: a straight motion toward the center of the Earth and circular inertia. But shared motion cannot be observed (we do not "see" the passenger sitting next to us in an airplane cruising at 600 miles per hour, even though we may know that he is). Thus we can perceive neither circular inertia nor the real (composite) motion of the stone. The normal observer only sees that component of the motion of the stone that he does not share: the motion toward the center of the earth. Therefore the stone does not fall straight down — it only seems to.

In this manner Galileo defused one of the main objections to the Copernican view. In doing so he pointed out that the crucial "facts" adduced by his opponents made theoretical assumptions; and then he advanced not a set of facts free of theoretical assumptions, and this is the main point, but a set of facts with different assumptions. The new real motions of objects were no longer directly observable, and the relativistic basis for holding this introduced a new way of doing physics. Different views of the universe, thus, lead to different assumptions, and different assumptions lead to different evaluations of what is to count as evidence. They also lead, as in the case of the Copernican revolution, to a profound transformation of our understanding of what the world is like.

Our problem of the justification of space exploration takes on an entirely new light. The conventional wisdom gave us a relatively static picture of science that makes it difficult to defend space science. If radiation from a certain region of the cosmos has been coming toward us for millions of years, and will be coming for millions more, what is the hurry to put a telescope in orbit to observe it right now? But if we place emphasis instead on the essential transformations of science and their consequences we gain a new point of attack.

2. *Scientific views determine what problems and opportunities we are aware of.* Since our world views tell us what the world is like, they also determine ultimately what opportunities we can take advantage of and what dangers we can be warned about. Thus with changes of world view comes the realization of many new opportunities and dangers.

3. *By becoming aware of new problems and opportunities, we also become able to think of new solutions and new technologies.* Einstein began his career by asking "useless" theoretical questions such as "What would the universe look like if I were traveling on a light ray?" In trying to satisfy his

curiosity about this and other equally impractical issues he was led eventually to develop his theory of relativity and to take a most decisive role in pushing physics toward quantum theory (although he later disagreed with the full-blown quantum physics of Bohr and Heisenberg). In these and other respects he changed several of our views of the world in profound ways, opening in the process the opportunity for a new understanding of light. This new understanding led to the theory of lasers. Lasers in turn opened up many technological opportunities. It was not long before some researcher decided to apply them in medicine. Today lasers are used in extremely delicate surgical procedures that would not be possible with any other technology known to medical practitioners. And it all began with a change of world view in a field far removed from medicine at the time.

In contrast imagine a crash program in Einstein's early days to have surgeons develop a surgical instrument that could do the sorts of things that a laser can do today. Is it reasonable to suppose that the point would have come when the well-funded surgeons would have realized that their aim required the overthrow of the physics of their day? And would they have then laid out the steps necessary to replace that physics with a view that would lead them to lasers, and so on? I think not. But without the new physics, could the surgeons have developed the equivalent of lasers, or of much other Western medical technology for that matter? Again I think not.

These considerations undermine the objection that all the good indirect results (e.g., the spinoffs) of the space program can be achieved by spending the money directly in the relevant areas. For benefits in one area may well require a prior theoretical transformation in another.

4. We may then conclude that serendipity is the natural (practically inevitable) result of scientific change.

Second Part of the Argument. Scientific exploration leads to scientific change.

Changes in world views are inevitable given the nature of science. For scientific theories — which are the underpinnings of our world views — are still our creations and thus imperfect. They are always in need of refinement, modification, or replacement altogether. The pressure for such changes comes from the double exposure to unusual circumstances (which force us to stretch our views) and to competing ideas (which are often developed to account for a few of those unusual circumstances, and then claim the right to extend to the entire field of the discipline). And, here is a key point, scientific exploration places science in new circumstances and presents it with new ideas. Thus scientific exploration leads to not merely the addition of a few, or even many, interesting facts but to the transformation, perhaps the radical transformation, of our views of the world.

Let me rephrase this second part of the argument in schematic form:

5. Science is dynamic. It must always be changing because
    (a) Science is never complete (being a human creation)

    (b) When science is challenged by new circumstances it must adapt (i.e., change).

    (c) When science is challenged by new ideas, it often has to change.

6. Scientific exploration places science in new circumstances and presents it with new ideas.

7. Therefore, scientific exploration leads to change in our scientific views.

Putting all these points together, we arrive at the conclusion of the whole argument:

8. *Since scientific exploration leads to change and scientific change leads to serendipity, exploration leads to serendipity.*

This argument undermines the notion that all the good indirect results (e.g., the spinoffs) of the space program can be achieved by spending the money directly in the relevant areas. For benefits in one area may well require a prior theoretical transformation in another. Supporters of exploration can now point out that the reason we can expect a bounty from space is precisely that the exploration of space presents many challenges to our science and our technology. Since space exploration is thus so likely to contribute to the transformation of our views, investing in it has a clear advantage over investing in fields not so ripe with challenge.

By showing that serendipity is an essential feature of science, this sketch of the nature of science allows the supporter of exploration to say with Descartes that a failure to explore is a failure to carry out our obligation to "procure the general good of mankind." The justification the supporter can now offer for exploration in general, and for the heart of space exploration in particular, begins to sound like a practical case. And to a large extent it is that, although the connections are more subtle and indirect and the time scale is larger than in the standard case. But it is a practical case born out of the nature of space science itself, and thus the guarantees that it offers go well beyond those of historical anecdotes.

This deeper and fundamental practicality forms the basis of the supporter's response to the strongest social objections. We must remember that the attempt to satisfy our scientific curiosity and sense of adventure through space exploration has resulted in those benefits discussed earlier. As those benefits become routine, the frontier of the unknown is pushed further back into the cosmos and our intellectual enthusiasm shifts accordingly. It has been the aim of my argument to show that in indulging our enthusiasm we are bound to force a change in our panorama of problems and opportunities. The supporter of exploration can now explain to the social critic why the previous benefits of exploration were not a fortunate accident: they were the result of the inevitable expansion of opportunity that is part and parcel of scientific exploration. Once we understand the dynamic nature of science we are in a position to vouch for its future serendipity. To a casual observer the heart of space exploration may not appear to have obvious practical benefits. But a deeper investigation reveals a long-term, fundamental practicality: the practicality that comes when a transformation of our views of the universe expands our range of opportunity.

This argument must still be defended against a variety of objections (e.g., on the *scientific* value of the space sciences) and it can use a good amount of elaboration. This task I have undertaken in a book-length manuscript, *The Dimming of Starlight: The Philosophy of Space Exploration* (forthcoming).

The argument works against the ideological objection as well. Unless we commit species suicide, we will continue to interact with the Earth and transform it in small and large ways. By doing so we act in the tradition of all other living beings: A tree grows tall and gives shade to violets and mushrooms that could not have lived without it; a beaver builds a dam, harming rosebushes and fish but helping water lilies and frogs; and once upon a time, bacteria gave the Earth its oxygen and nitrogen atmosphere. The question for us is not of whether we will interfere, but rather of how much and how wisely. Now, to act wisely we need knowledge of the Earth, of us, and of our interactions with the Earth. Otherwise we are likely to impose too big a burden on the planet or on its human inhabitants. That knowledge, that human knowledge is not complete as of this writing — not even environmental activists can reasonably claim that they know everything about our planet. Thus the same considerations about scientific knowledge discussed earlier in this paper apply to knowledge of the Earth. Once again, we need a dynamic science that can change our panorama of problems and opportunities. To deprive ourselves of it is to deprive ourselves of the chance to act wisely. We are already a big presence on the Earth and need to move carefully. It would thus be irresponsible to forego the tools that may illuminate our way (tools such as the Mission to Earth). Space exploration is thus not a false panacea but an important means to a cleaner and better future.

References

Clark, Wilson (1977): *Space Colonies*, Brand. S., ed., Penguin Books, p. 36.
Clarke, Arthur C. (1946): "The Challenge of the Spaceship," *Journal of the British Interplanetary Society*, p. 68.
Feyerabend, Paul K. (1975): *Against Method*, NLB.
Greve, T., Lied, F., and Tandberg, E., eds. (1976): *The Impact of Space Science on Mankind*, Plenum Press, p. 13.
Holman, Mary (1974): *The Political Economy of the Space Program*, Pacific Books.
Kuhn, Thomas S. (1970): *The Structure of Scientific Revolutions,* University of Chicago Press (2nd. Edition).
Lakatos, Imre (1978): *The Methodology of Scientific Research Programmes*, Cambridge University Press.
Popper, Karl R. (1959): *The Logic of Scientific Discovery*, Hutchinson.

14 Pecking Orders and Rhetoric in Science

There is a pecking order in the natural sciences, with physics clearly at the top. This is so for three reasons. One is historical: physics led the way in the scientific revolution and presumably set the standards for subsequent science. A second reason is that physics deals with processes that are fundamental to all the natural world; it deals with what all objects have in common. Important changes in physics tend to be felt in many scientific places. The third reason is that the mathematical and experimental rigor of physics, coupled with extraordinary feats of the imagination, maintains the prestige of the discipline very high.

Even if such high prestige is well deserved, there is an unfortunate tendency to think that other sciences have much to learn from physics but little to teach it. And even within physics we permit a pecking order with the "hottest" areas at the top. A result of this situation is the expectation that only scientific expertise can properly decide what research should be given the highest priority. These are supposed to be internal disputes. After all, experts ought to be the most capable judges of the worth, of the track record, and of the promise of the competing disciplines and approaches — at least in pure science. In applied research, of course, priorities are often dictated by social, political, and economic exigencies — to no one's surprise.

I think that the testimony of experts on these matters should, indeed, receive much weight. But I also think that such testimony suffers from great limitations in a context that is, to a large extent, rhetorical. In other words, considerations drawn from the successful science of the time alone will turn out to be too narrow in estimating how the balance of fundamental research may shift between different areas. Even within physics itself, larger considerations — some from fields as remote as the philosophy of science — need to be brought together in a rhetorical fashion to defend the support of one area of research over others. In this short paper, I cannot attempt a full argument for this thesis, but I do intend to illustrate it by studying a contemporary case within physics: the shift in the perception of space science as fundamental research.[1]

For several decades now, the most pressing problems in physics have been thought to reside in the very small, for the simple reason that the smallest components of matter are presumed to be the building blocks of the universe. These problems are the province of particle physics, and within particle physics the leading view used to go by the strange name of chromodynamics, later replaced by string theory, which in turn seems to have been abandoned. The central notion of chromodynamics was that quarks can be used to explain many of the most important properties of subatomic particles.

About the time when these ideas were beginning to take particle physics in new directions, in the early 1960s, space exploration was also beginning to be done in earnest. In those days it was not uncommon to hear scientists scoff at the possibilities that space exploration offered for fundamental research.[2] Since particle physics must be carried out largely in very expensive particle accelerators, the suspicion arose in some minds that the also expensive space science was taking money away from the cutting edge of physics. Not that particle physics was presumed to be the only kind of research that a respectable, physicist could undertake. But other physics shines by reflected glory, so to speak, and thus the more removed from the center of the discipline, the less important it is thought to be.[3]

This uneasy feeling about space science has been losing ground all the time, and for very good reasons. The importance of quarks is not merely that they presumably explain the variety and properties of sub-atomic particles. Quarks also point the way toward a unified account of the basic forces that act between particles, which physicists often call the "basic" forces of nature. Fundamental research is then research about those forces, and that is the research done at the gigantic particle accelerators.

Those basic forces of nature are the electromagnetic, the weak, the strong, and the gravitational. Current physical theory has developed a more or less unified account of the first two, and there is hope that the strong nuclear force can be explained along the lines of investigation laid out by particle physicists. But gravity is not so tractable. The desire for the unification of physics has brought forth many so-called "supergravity" theories. And in this case, the name is very telling. The best hope for a unified account of all the forces requires the description of a state of the universe in which all the forces are of comparable strength. But today gravity is far weaker.[4] As far as the current thinking goes, we must study the beginning of the universe to find a clear case of such strong gravity, "supergravity," in interaction with the other forces.

Particle physics, then, points the way to *cosmogony*, the science of the origin of the universe. But how do we study such origin? With the tools that astronomy and astrophysics make available to us. Such study cannot, of course, be separated from the study of the nature of the universe; that is, it cannot be separated from *cosmology*. And here, space exploration becomes of the greatest importance, for it has opened to us opportunities for cosmological research that are as necessary for physics as they are impossible to carry out on the surface of the Earth.

I will consider two of those opportunities. The first is very simple. Doing cosmology from the Earth is very limited because most of the information that comes from the cosmos is blocked by the atmosphere. Observatories placed beyond the atmosphere eliminate that obstacle and permit us to examine the universe in the entire range of the electromagnetic spectrum. Different physical phenomena, different interactions of matter and energy, produce different kinds of effects, and those effects involve different bands of the spectrum. Therefore, the change that is brought about by science done from space is not a mere increase in the number of cases that we examine, but a rather profound qualitative change in what we may examine.

The point is that fundamental physics requires cosmogony, that cosmogony requires cosmology, and that cosmology cannot be done properly without placing telescopes in space. I will expand these remarks presently, but first I would like to place them in an historical context. For the relationship between concerns about the nature of reality has characteristically been tied to concerns about the nature of the heavens. This should be borne in mind when we realize that many phenomena of crucial interest can best be examined at those frequencies absorbed by the Earth's atmosphere. They are of crucial interest precisely in that they offer essential clues as to the origin and evolution of the universe. What space exploration has come to offer, then, is the opportunity for a radical transformation of our most fundamental theories of nature. Astrophysics may once more lead the way, as it did during the Copernican revolution.

When Galileo turned his telescope to the heavens, a new sense was born. But this was no mere addition, for Galileo's telescope, in giving us tens of thousands of new stars, the moons of Jupiter, and the phases of Venus, helped usher in a view that was contradicted by direct experience. As Galileo himself said, "there is no limit to my astonishment, when I reflect that Aristarchus and Copernicus were able to make reason so to conquer sense that, in defiance of the latter, the former became mistress of their belief."[5] But in the telescope he found a "superior and better sense than natural and common sense." This new sense, furthermore, was not just a refinement of sight, but rather an alternative that agreed with the Copernican view, unlike plain sight. Indeed, whereas to plain sight the magnitude of Mars changed little — a troublesome fact for the Copernican view, according to which the distance between Mars and the Earth varied considerably — in looking through the telescope, the magnitude behaved as if God had been a Copernican.

In this way, a new technology came to the rescue of an idea that was to transform our view of nature most profoundly. Not that the telescope was free from reasonable question, for on the contrary its reliability was a lucky assumption made plausible more by Galileo's enthusiasm than by his argument. But the telescope did open up many avenues of observation and investigation that would not have been there otherwise. It was a promise from the heavens, better kept in the course of the new science than perhaps Galileo had a right to imagine. For him, it was too striking a coincidence that the telescope would so match the new astronomy of Copernicus. For others whose fundamental views were at stake, it was a case of a distorting instrument of observation presuming to support a refuted and obnoxious view of the cosmos.

To most of us now, it seems fortunate that, as Galileo put it, Copernicus "with reason [theory] as his guide ... resolutely continue to affirm what sensible experience seemed to contradict."[6] Many of those who were in a position to choose, decided in favor of the most exciting of the alternatives (in accordance with a principle that Brian Toon calls "Sagan's Razor"). Whether they had more compelling reasons I shall not discuss here. Suffice it to say that the invention by Newton of the reflecting telescope, and the refinements of both kinds of telescopes, permitted not only the discovery of many new objects in the universe, but also a shift in perspective about its nature. And having embarked on a different approach to nature, the new scientist also had the motivation to develop the auxiliary sciences that eventually established the reliability of the telescope as well.

Such reliability is rather limited, given what we now know about the interaction between electromagnetic radiation and the Earth's atmosphere. Galileo's new sense, our small window into the universe, became better and better, and when it seemed that it was reaching its limitations, in the middle of this century, we discovered a new window: radio astronomy; and then another: infra-red astronomy. Much resulted from the evolution of Galileo's instrument. But then with the discovery of radio- and infra-red astronomy, new kind of objects, and new kinds of activity in objects already known, gave us a glimpse of what a look at the universe in the full electromagnetic spectrum may do. Quasars and pulsars, and the radiation left over from the big bang, began to tell us about a universe far stranger than we had imagined. With the advent of space exploration, we can now examine the universe in the ultraviolet, x-ray, and gamma-ray regions of the spectrum as well. Moreover, we can also enhance the previous means of astronomical research, since now all the bands of the infra-red, for example, are open to exploration. And even in the visible range, a space telescope can overcome many of the handicaps of its terrestrial counterparts — atmospheric refraction, background lights, daylight, bad weather, columns of air inside the telescope, gravity effects on the shape of the lenses, and many others. Although a new generation of Earth-based telescopes will reduce some of these problems, they will be hard put to match a performance that is superior by orders of magnitude — being able to sweep a volume 350 times larger and produce images ten times finer. We can only speculate as to what the evolution of such space telescopes may bring.

The x-ray and gamma-ray observatories will permit us to investigate phenomena such as neutron stars and the vicinity of black holes, where the violence of massive gravitational collapse surpasses anything we have ever observed. It is not plausible to say that we have a complete idea of the nature of the universe while leaving out phenomena which obviously may have great consequences for our theories of gravitation. And it is not plausible, either, to fasten upon certain ideas of the origins of the universe without the guidance of reliable theories of gravitation. Now, since unification schemes must deal with questions of the origin and evolution of the universe, space astronomy is clearly fundamental science by even the narrowest and strictest of criteria.

Of course, there is the possibility that all present unification schemes are entirely misguided. But if that is so, space astronomy will be particularly helpful in pointing to areas of physics where new directions would be fruitful. First of all because without the boost of space astronomy to cosmology, our models of the evolution of the universe will remain far too speculative to determine the strengths and weaknesses of any potential unification of the basic forces of physics. This illustrates a second thesis for which I will argue below: that space science helps transform theoretical into experimental science. Science needs to rub against the world, for such friction polishes and sharpens the rough guesses that humans make about the universe.

Furthermore, this possibility of studying black holes and other strange objects presents extraordinary opportunities to challenge all of physics. It has been said that in black holes, physics comes to an end. The reason is that in a runaway gravitational collapse, which is presumably what exists in a black hole, matter and energy disappear into a single geometrical point at the center of the black hole. This point, called a "singularity," obviously contains no space. And it has not time either since time slows down in the presence of a strong gravitational field. Where the field is practically infinite, time simply does not "happen." But the laws of physics make no sense outside of time and space. Thus we have a situation in which matter-energy is no longer subject to the laws of physics as we can conceivably understand them. These are strange possibilities to loosen the grip of entrenched ideas.

With the advent of space exploration, thus, the study of gravity is likely to regain its historical importance. This is as it should be: Apart from being one of the fundamental forces of nature, gravity is pervasive throughout the universe, and critical to many interesting phenomena that may hold the key to our understanding it much better. The decline of its popularity as an area of research, in the middle of this century, was probably the result of the lack of opportunity for experimentation in Einstein's general theory of relativity, the most accepted view of gravitation. For a long time, the only test of this theory were the shifting of starlight during eclipses and the perihelion of Mercury. A third test, the red shift of radiation in strong gravitational fields, was not entirely convincing until the work of Alley, which measured the change of time in clocks flown in airplanes at varying altitudes, an experiment designed with space technology in mind, and the suborbital flights of maser clocks in Vessot and Levine's experiments (circa 1972).[7]

In this matter of turning the general theory of relativity into an experimental science, the first contribution of space exploration has been the proliferation of tests of general relativity itself, and even of alternative theories of gravity. Sometimes this experimental work very strongly complements previous avenues of investigation. We can see examples beginning in 1965 with the delay of radar signals bounced off objects going around the sun, which complemented the eclipse test, and continuing today with the search for gravity waves using spacecraft. The importance of gravity waves, were they to be found, is that matter is transparent to them. That is, we can use them to probe the universe in ways not avoidable, even with the full electromagnetic spectrum because matter, after all, absorbs, refracts, delays, and reroutes electromagnetic radiation.

Sometimes this experimental work is not only entirely new, but can be done only in space. Another of Alley's experiments, bouncing laser beams off the Moon to set a test between Einstein's general relativity and Dick's theory of gravitation, depended on reflectors that were placed on the Moon by American astronauts and Russian robot lunar landers. But the most fascinating experiments are yet to come. Of particular significance is the orbiting gyro experiment directed by Francis Everitt at Stanford University. This experiment, scheduled to be launched by the Shuttle in the late 1980s, will test the general theory of relativity in two most significant respects. According to Einstein's view, the distribution of mass determines the geometry of space-time. But when a big mass rotates, it places space-time in motion as well, that is, it drags space-time along. To make a very precise calculation of the value of the drag, we have to know, very precisely, the values of the mass and the rotation involved. This we know of the Earth, which also has a mass big enough to permit, according to the theory, a small but detectable amount of space-time drag. A small, nearly perfect sphere — a gyroscope — placed in polar orbit will have its rotational motion affected by such a drag. The object of the experiment is to measure that effect on the rotation — more specifically, the precession of the sphere with respect to the fixed stars — and see how it agrees with the values predicted by the theory.

Whether this experiment agrees with Einstein or not, it will be seen as a critical examination of a fundamental block in our understanding of nature. But the experiment has a double purpose. It also aims to measure a most startling prediction of the general theory of relativity. When a electrically charged body rotates, it produces a magnetic field. Because of the structure of general relativity, we can ascribe a similar effect to rotating gravitational masses. That is, a field should be produced that would create torques on the orbiting gyroscope (this would lead to what is called the "gravito-magnetic precession of the gyroscope"). A positive result of this experiment would confirm the existence of magnetic-like properties of gravity, a discovery of the most extraordinary importance. The precision required in this experiment is so incredibly high that, if it were carried on the Earth, the distortions introduced by the Earth's gravity would make it practically impossible.[8]

These are just two of the aspects that today place space physics and space astronomy at the cutting edge of fundamental research in physics. They could have been used in the initial debate about the value of the exploration of space, though in a much vaguer form than it is possible today (the unification of the basic forces of nature was not a widely accepted goal of physics in those days, for example). And to some extent they were so used. They were not likely to effect immediately a profound change of expert opinion, precisely because the promise of space was not as clear and explicit as the most ordinary research possible in particle physics. But their rhetorical value proved decisive with NASA and other space agencies which, of course, had a vested interest in seeing that space exploration should yield significant science.

There is no offense to reason in permitting rhetoric a role in disputes about scientific value and promise. If we convince ourselves that the matter ought to be decided by some rigid canons of rationality (which, of course, experts are best qualified to apply), then we risk taking too narrow an approach where daring may be called for. I do not wish to suggest that this rhetorical option can be used by people without any acquaintance with the field, for I suppose that they are not likely to know what moves are available to undermine the expert testimony. In the case that I have examined, those moves included standing back from the current practice of physics to permit some larger considerations.

One of those considerations is simply that to the extent that microphysics (i.e., particle physics in this case) determines what processes underlie the macro-phenomena of the large universe, then the study of those phenomena ought to serve as an independent testing ground for our theories of the small. Another consideration is that scientific views are ways of interacting with the universe. As such views are forced to encounter new kinds of situations, they are going to change, and in changing they will affect the structure of the field. This is a point from the philosophy of science, and its discussion deserves a much larger work. But I suspect in any event that its implicit recognition made the rhetorical appeal of the argument for physics in space so much the greater. For surely space exploration offered, if nothing else, the opportunity to deal with entirely new kinds of situations. It is fortunate that NASA had a vested interest in the matter.

Notes

1. A fuller treatment of the issue can be found in my upcoming book, *The Dimming of Starlight: The Philosophy of Space Exploration*. Some passages in this essay are excerpts from that work.
2. In a survey done in 1964 by *Science*, the Journal of the American Association for the Advancement of Science, 16% of the science Ph.D's who responded, agreed with President Kennedy's decision to go to the Moon, while an overwhelming 64% disagreed. Even today, in giving talks about the nature of space science, I am often confronted with the objection that if science is our goal we should spend the money on particle physics instead.
3. For an account of the low status suffered by the planetary sciences until rather recently, see Stephen G. Brush, "Planetary Science: From Underground to Underdog," *Scientia*, 1978, vol. 113, p. 771. Brush demonstrates how the prejudice against planetary science was completely blind to the history of physics.
4. Taking the value of the strong force to be 1, the other values are as follows: electromagnetic, 10^{-2} ; weak, 10^{-6} ; gravity, 10^{-40}.
5. Galileo Galilei, *Dialogue Concerning the two Chief World Systems*, Berkeley, 1953, p. 328.
6. *Ibid.* p. 335.
7. For an account see Carroll O. Alley, "Proper Time Experiments in Gravitational Fields with Atomic Clocks, Aircraft, and Laser Light Pulses," a lecture published in *Quantum Optics, Experimental Gravitation, and Measurement Theory*, P. Meystre and M. O. Scully, eds., Plenum Publishing Co., New York: 1982.
8. To achieve the desired performance in the experiment in orbit, the deviation in the radius of the sphere must be less than 10^{-16} of the radius which, in this case, means that the deviation must be less than 1/1000 of a *nuclear diameter*. That is simply out of the question. For details on the experiment, the reader may consult C.W.F. Everitt, "Gravitation, Relativity and Precise Experimentation," in *Proceedings of the First Marcel Grossman Meeting on General Relativity*, Remo Ruffini, ed., North Holland: 1978.

15 SETI, Self-Reproducing Machines, and Impossibility Proofs

The recent discovery of planets around other stars and of organic carbon in a Martian meteorite has renewed scientific interest in the search for extraterrestrial intelligence (SETI). I intend to examine one important argument against SETI that touches on some issues from the philosophy of science, particularly of AI and of biology.

Inspired by Copernicus, *proponents* of SETI assume the *principle of mediocrity*, which asserts that the sun is a typical star in having a planet like the Earth, propitious to the origin of life, that terrestrial life is typical in having produced intelligence, and that human intelligence is typical in giving rise to a technological civilization.[1] The *opponents* of SETI stretch this principle slightly to add that a technological civilization is typically expansionist. As a result they are able to produce a variety of "impossibility proofs" against the existence of extraterrestrial intelligence (ETI). Although the principle of mediocrity itself is in need of philosophical criticism, I will restrict my remarks to the main impossibility proof.

If ETIs do exist, Fermi once asked, why aren't they here? The argument is, briefly, that if the SETI proponents are right, there should be technological civilizations far older, and presumably far more advanced than ours; for many "typical" stars in our galaxy are billions of years older than the sun, and thus in some of their planets intelligence should have sprung long before it did on Earth. Now, just as we expanded from our beginnings in Africa, any civilization capable of space flight is bound to expand throughout the galaxy. Furthermore, this expansion would take place in a short amount of time (travelling at .1c, which is within human reach now, it would take only one million years to cross the galaxy). Thus the ETIs should be everywhere in the galaxy by now. But they clearly are not here; therefore there are no extraterrestrials in this galaxy.[2]

These opponents realize that interstellar travel may be too long and arduous for biological beings. They also realize that it may be unfeasible, even for advanced civilizations, to send "unmanned" probes to the — possibly — billions of planets in the galaxy. Their impossibility proof depends instead on a technology they believe is inevitable: self-reproducing machines.

John von Neumann supposedly proved already that a machine could be designed to make copies of itself.[3] A more advanced civilization would surely have discovered the equivalent of von Neumann's proof, and would be in a position to develop the appropriate technology. Indeed, some NASA scientists are investigating the possibility of using such machines to explore the galaxy.[4] We already have a mathematical proof that self-replicating machines can exist, all we need is the talent, effort, and money to create the technology. If this seems feasible for us, an advanced technological civilization surely would have no trouble building a couple. Once they arrive at their destinations, these machines would make copies of themselves, which would then move on to the nearest stars, and so on, setting in motion a geometric progression, until they overrun the entire galaxy. But we have no evidence of such machines here; therefore no advanced civilizations exist.

Von Neumann offered not one but five proofs. The first one, however, is the main basis on which all these promising speculations rest. Von Neumann knew that, through evolution, organisms produce others more complicated than themselves, and he wondered whether the opposite had to happen with machines. So he set out to determine whether it was possible to program a computer to make a copy of itself. He imagined a robot floating in a vat full of robot parts. The robot could be programmed to pick up a part and identify it. Then the robot, which had a blueprint of itself, would look for the connecting parts, and would then begin putting another robot together the way a child puts together a Mechano set. Surely, we can also program a computer to do this (it is far more complicated, but possible). Once the second robot was assembled, the first would pass on to it the self-replicating program (or set of programs, rather). By breaking down the task of self-replication into small, manageable tasks, von Neumann thought, an automaton could copy itself. This result led him to remark that there were two kinds of automata: artificial automata, such as computers, and natural automata, such as people and cats.

Thus the implications of this very simple conceptual (rather than mathematical) proof go well beyond the concerns of technology and exploration — they affect also our notion of life. But let us see what happens when we send one of von Neumann's self-replicating automata (SRAs) into the galaxy.

The first thing that comes to mind is that when an SRA gets to another world it is not going to land in a vat full of parts. It will have to build factories to build the parts from the raw materials that it will mine. But the factories are themselves made of parts, so it will have to build other factories to build the parts to build the factories. . . .This is called the "closure problem" by those in the field. There is no need to fear an infinite regress, however, for we know that the closure problem can be solved — even if we still have no idea how to program a machine to solve it. And we know that it can be solved because our technological civilization solves it: we do send rockets to other worlds.

We must realize, however, that such an SRA will be an extremely complex machine, both in its computer programs and in its physical realization. Indeed, an SRA will be the equivalent of a technological civilization (including the starship by which it moves about). Whether we can write a program that complex, and whether we can assemble and then make to work a machine that sophisticated is open to question. But let me assume for the sake of argument that we can.

Let us imagine what would happen when one of these extremely complex SRAs arrives at a planetary system. Now, we do not yet know what other planetary systems look like, but some theories suggest that they would be collections of small rocky planets and gas giants. Let us suppose now that an SRA comes into *our* solar system. Jupiter and Saturn would not be good places to land (even figuratively) because it is unlikely that the SRA can fashion the needed parts and factories out of the hydrogen, helium, and the trace gases that can be found in their atmospheres. The moons of the gas giants are not that much better, for surely a machine equivalent to a technological civilization may be presumed to need a variety of materials, including metals, for, the task of self-replication. Unfortunately the low density of most of these moons (less than 2.0 g/cm^3) suggests that they would not be good places to search for the necessary raw materials.

Nevertheless we know that in rocky planets like Earth an SRA can find practically everything it needs (or so we hope). But even on rocky planets the SRA's problems are far from over. Small differences between the planets in astrophysical terms may lead to significant differences in density and chemical composition of the atmosphere. These significant differences will in turn make it necessary to adopt different strategies for mining and manufacturing. For example, on Earth the best way to treat some particular ore may be to throw it into a pot of boiling water. In Mars the water would evaporate before the ore is settled in. In Venus the pot itself might melt.

This is by no means a small problem. No matter how similar planetary systems may be to one another, we should still expect at least some small astrophysical differences between their rocky planets. Thus the possible combinations of factors that may affect mining and manufacturing may be practically infinite. Therefore the already extremely complex SRA would need, in addition, some general purpose programs so it can begin the task of making the parts for its progeny. But no one knows how to write such a general purpose program, and there are reasons for thinking that they cannot be written.[5] The biggest stumbling block to artificial intelligence has been precisely the inability to write programs that exhibit a flexible response to run-of-the-mill environments, let alone to the incredible variety demanded of SRAs. Nor is there any assurance that this problem can be overcome in the foreseeable future. (Connectionist approaches, which present a significant alternative to von Neumann's view, are better able to handle context, but this situation would still present too tall an order for them.)[6]

But if rocky planets are too heterogeneous for the SRAs' needs, we may still find a homogeneous environment where they can get all the raw materials in question: the asteroid belt. We may be stretching our luck by supposing that all planetary systems (or even most of them) have asteroid belts, but let that pass.

An SRA could move from asteroid to asteroid picking up metal ores here, carbon compounds there, mining and processing them all in a rather stable environment (the cold vacuum of interplanetary space — although the exact location of the asteroid belt, and the strength of the stellar wind in that system may again provide too much variety). After granting all this, we are now able to deal with the fundamental problem of SRAs. As von Neumann himself pointed out, the more complex a computer program is, the more likely it is to have errors. But the SRAs would be far more complex than anything we have ever imagined programming and building. These errors, furthermore, involve principally the task of self-replication. I am talking not only about bugs in the gigantic program but about errors of execution in manufacturing and assembling the many components (an alloy that is not quite up to strength, a tooth in a gear that is slightly short and with a bit of wear will no longer catch another as it must).

Neither quality control nor error-detecting programs will solve this problem, for it takes only a small percentage of error to bring the task of self-replication to a halt. Let me explain the difficulty by means of an illustration. The SRA is already saddled with a computer program so complex that it seems difficult to imagine that we can debug it completely. But now we must add to it a program that must equip the machine with ways of checking the complete specifications for all parts, all fittings, and all functions. Nonetheless even this added complexity does not solve the problem. For a program that can foresee all the possible ways in which something can go wrong (a piece of dust, a screw partly loose) begins to look like a general purpose program (that is one reason why astronauts are so useful in space). In a machine that complex, engaged in the extraordinarily complex task of producing another SRA, things can go wrong in more ways than we can imagine. A program that must deal with so many unknown contingencies is, again, a program that can deal with an open-ended environment. And that is where SRAs come to grief, once again.

But isn't it the case that living things, which are very complex in their own right, also make errors in the copying of the information used in replication? So why is it that error *must* bring machine replication to a halt when it does not do so in the replication of living things? The answer is as follows. In living things "errors" (mutations, recombinations) do serve to provide genetic variation in a population. Actually, many mutations are supposed to be deadly, but in some cases the genetic variation allows the population to adapt to changes in the environment; that is, as the environment changes, some of the members of the population may take advantage of past "errors," and the population lives on. This is how error can be adaptive for living things.

In the case of the SRAs, however, we must remember that the asteroid environment was chosen precisely because it would not change from one system to another. In that unchanging environment there is no advantage in error. In SRAs the bulk of the errors that concern us here are precisely in the reproductive part, and thus they are maladaptive. When all is said and done, it seems that an SRA technology is not really even a gleam in a scientist's eye.

Nevertheless many would think that we can actually point to examples of self-replicating machines: trees, cats, humans. These people already assume von Neumann's conclusion, that there are two kinds of automata — natural and artificial. I suspect that they find that assumption reasonable because they believe that the genetic code is the equivalent of a computer program, and thus they conclude that living things are just the realizations, or executions, of their particular programs. This view is no longer popular amongst biologists, but we still need to see why it fails to help the case of the SRAs.

First of all, if we are to use analogies, we should stress the following: in the case of the SRAs, the machine must make a copy of itself and then pass on the program (even if it passes on the program to a unit that is not yet completed, the point is that making the copy and passing on the program are separate, largely independent tasks); in the case of living things, however, it seems more proper to say that they pass on the program first, and then, as result of that, the copy is made. This is not a small difference, for in living things relatively simple accomplishments (e.g., a fertilized egg) can produce very complex organisms (the egg is extremely complex chemically, but simple relative to, say, the human being that will result from it).

This result leads to a second, and more important point. To picture the genetic code as a computer program is just to engage in metaphor, and the metaphor is highly misleading. The "instructions" of the DNA produce the expected results (e.g., proteins, cells, tissues, organs, behavior) only because at every level they can be expressed in appropriate environments, indeed it is often the appropriate environment that will trigger the next stage in embryonic development. In a human, for example, at a certain time in the life of the embryo the normal development requires a certain concentration of sodium, and after the human is born, the attention paid to him is not only necessary for his survival but provides the stimulation needed for the central nervous system to grow further.

In other words, the "instructions" of the DNA do not have meaning by themselves. This issue is similar to that of the meaning of words in the philosophy of language. It used to be thought that words had intrinsic meaning, but it is generally accepted now that the meaning of a word depends just as much on the context in which it is uttered (the meaning is given by the interaction of word and context, where the context may include a large variety of factors, including the relationship of the word to other elements of the sentence, the manner of its utterance, the social conditions that the speaker and the listeners take themselves to be in, etc.). In embryonic development, the "program" makes sense (has any meaning) only in that the maturing organisms interacts with a sequence of appropriate environments. Those environments provide the biological contexts in which the "instructions" of the genetic code are instructions at all.

Now, that sequence is itself the result of natural history, that is, of a long series of interactions between the ancestors of that organism and the environments in which they evolved. There is a clear sense, then, in which a living being comes into a world that is already made for it. The world of the SRA, on the other hand, must be largely described in its program from the beginning. The meaning of the program must be made explicit beforehand. To illustrate this point, let us consider the development of a nervous system. In dissecting an animal we may find that its nerve cells always exhibit a certain pattern, and may thus imagine that pattern is contained in a blueprint in the DNA. Nevertheless, as the nerve cells grow through, say, a muscle tissue, they need not be guided by any such blueprint. They may simply have "instructions" to grow in the general direction of a chemical marker, until they make contact with a membrane, which will turn the "instructions" off. But the developing muscle cells will then constrain the manner in which the nerve cells grow (the nerve cells will have to grow around them, for example). The final pattern is the result of such contingencies, and there is no need for any blueprint whatsoever.

In living things the burden of development is largely assumed by natural history, in SRAs it mostly falls on the programmer's shoulders. That is why the first is manageable and the second is not. This is not to say that artificial life is impossible. As far as I can tell, there are no objections in principle against it. But it would be *life* nonetheless. Insofar as there is design in it, that design is grafted on to the knowledge we have of natural history, to take advantage of prior interactions with environments or sequences of environments. (I am not referring here to the computer field of "artificial life," based on Von Neumann's other proofs, which has conceptual problems of its own.)

I conclude that there is no good reason for thinking that a technology of self-reproducing machines is possible, much less practically inevitable. The impossibility proof fails.

Notes

1. The classic expositor was Carl Sagan, see for example his (ed.) *Communication with Extraterrestrial Intelligence*, MIT Press, 1973.
2. Authors more contemporary than Fermi have developed the argument criticized here. See for example, Frank J. Tipler, "Extraterrestrial Beings do not Exist," Physics Today, April 1981, pp. 9-38. For recent commentary (and the reference to Fermi) see Paul Davies, *Are We Alone? Philosophical Implications of the Discovery of Extraterrestrial Life*, Basic Books, 1995, based on his series of lectures at the University of Milan in 1993.
3. John von Neumann, *Theory of Self-Reproducing Automata*, A.W. Burks, ed., University of Illinois Press, 1966.
4. *Advanced Automation for Space Missions*, NASA Conference Publication 2255, 1982.
5. The classic critique is Richard Dreyfus, *What Computers Can't Do*, Harper & Row, 1972.
6. Even in optimistic treatments such as Paul M. Churchland's, *A Neurocomputational Perspective: The Nature of Mind and the Structure of Science*, MIT Press, 1989. I find Churchland's view very plausible, and in a sense this paper supports it by undermining the view of mind (and body) put forward by von Neumann.

APPENDICES ON FEYERABEND

Appendix A
Science in Feyerabend's Free Society

Paul Feyerabend's latest book, *Science in a Free Society*, is certain to intrigue some, appall others, and infuriate many[1]. In a free society, he claims, there should be a separation of science and state, just as there is a separation of church and state. This should be so because (1) science is only one of many traditions or ideologies, (2) a free society is one in which there is equality among traditions or ideologies, and (3) society's preferred treatment of science in education, medicine, and so on (making it the "fabric of democracy") violates the rights of other traditions. A central concern of the book is whether modern science can be shown to be better than alternatives such as magic or Aristotelian science. Feyerabend argues that it cannot. But there is much more to the book: Feyerabend examines the relation between reason and practice, tames the specter of relativism conjured up by his views, attacks modern medicine, and spends a good many pages debunking experts and exposing the low caliber of contemporary philosophy of science. Just as in his *Against Method (AM)*, one can find " ... arguments to make rationalists feel at home, arias in various keys to please the more dramatic reader, fairy tales to capture the Romantic, ... rhetoric for those who like a hard-hitting debate with no holds barred, ... personal remarks for people who rightly feel that ideas are made by men and that one understands them better more one knows about the minds that create them"(180). Feyerabend disappoints neither friend nor foe.

For all of Feyerabend's clarification of his position, there is no guarantee that *Science in a Free Society (SFS)* will escape the misunderstandings that greeted *Against Method*. At the risk of adding a few more misunderstandings of my own, I would like to sketch and criticize what I take to be Feyerabend's main theses in his latest effort — a risk not lightly taken, by the way, after considering his devastating response to critics in "Conversations with Illiterates," the third and longest section of the book. I will argue in particular, that there should be no separation of science and state and that science is, indeed, the basis for the "fabric" of a free society. This is not to say that calm should return to the world of philosophy of science, for Feyerabend's work forces a reappraisal of the most cherished standard positions even if it does not succeed completely.

I think it is not objectionable to think of science as a tradition. But some may find fault with putting science on par with other traditions. Science is said to be a rational enterprise; witchcraft and astrology are not. Thus, science is a privileged tradition if it is a tradition at all. Feyerabend is not impressed, however, and so he mounts an attack on two fronts. First, he notices that people who make such claims tend to think of rationality in terms of rules or standards. To be rational is to live up to the standards, that is, to follow a *methodology* (a collection of rules such as "reject hypotheses that conflict with the facts," "avoid *ad hoc* moves," and so on). Science is supposed to be successful because it is rational, and rational because it has the right method. Against this position, Feyerabend sketches again the arguments from AM. There, he pointed out that in crucial episodes in the history of science — in exemplary incidents of scientific success or progress (a) the most basic rules were violated, and (b) they *had* to be violated if progress was to result. Therefore, sticking to a methodology would have had the most disastrous consequences (from the rationalist's point of view — in this as in the nature of progress, Feyerabend accepts the rationalist account for the sake of argument only). So the rationalist has a most unwelcome choice: science is either progressive or rational, but not both. This is not to say that all rules are to be abandoned, for Feyerabend points out that some procedures actually aided the scientists in particular circumstances. His slogan "anything goes" is simply the description of the situation from the rationalist point of view and not a new methodological proposal. It should be an embarrassment to the profession that many reviewers were completely unable to see the structure of this simple reduction. In *SFS* Feyerabend repeats and clarifies his arguments against a universal scientific reason or methodology, and expands his case against contextual variations of it (pp. 164-166).

Not all methodologies are of the sort described, however. Lakatos' methodology, for example, does not demand the application of such standards but requires, instead, progressive research programs in the long run. But even a research program that is stagnating cannot be eliminated, for nothing rules out its comeback (a progressive phase) if given time to regroup. The problem is that Lakatos' standards do not provide a measure of how much time is enough. Thus, Feyerabend argues, no proposal can ever be eliminated, and Lakatos' standards are therefore empty, mere verbal ornaments.

Suppose, nevertheless, that it could be shown first that science is rational. Would it follow then that science should be given preference over other traditions? No, says Feyerabend. It must be shown first that being rational (in the required sense) is best. To show that it is not best — in spite of the apparently overwhelming obviousness of the claim — or at least that it has not been shown to be, Feyerabend makes use mainly of a general argument and a particular test case. The general argument is that the distinction between reason and practice is illusory. The alleged conflict between these two is rather a clash of traditions, for reason is nothing more than another tradition. But traditions are neither good nor bad; they simply are. Thus, rationality cannot serve as an "objective" arbiter of traditions, for it is itself a tradition. Of course, traditions may be made to serve as arbiters of disputes anyway, but then the whole affair is relativistic, surely not the sort to gladden a rationalist's heart. In a particular case Feyerabend considers whether it can be shown that modern science is better than Aristotelian science. The answer, again, does not provide much comfort to the defenders of science ("Aristotle Not a Dead Dog").

An immediate consequence of Feyerabend's general argument is relativism. Not to worry, he says. Humanity will not be any the worse for it. This matter may be of the greatest importance. But I would like to concentrate first on Feyerabend's case for it and on the support that case provides for the thesis of the separation of science and state.

To begin, Feyerabend argues that claims to objectivity assume a distinction between traditions (practices) and a different domain that is outside of all traditions but may act on them (20). Feyerabend proceeds to an historical analysis of this distinction. It arises, he says, because of "the tendency to view differences in the structures of traditions (complex and opaque vs. simple and clear) as differences in kind (real vs. imperfect realization of it)" (22). Thus, for example, "In antiquity the relation between the new entities of mathematics and the familiar world of common sense gave rise to various theories. One of them which one might call *Platonism* assumes that the new entities are real while the entities of common sense are but their imperfect copies. Another theory, due to the *Sophists*, regards natural objects as real and the objects of mathematics (the objects of 'reason') as simple minded and unrealistic images of them"(21). This tendency is "reinforced by the fact that the critics of a practice take an observer's position with respect to it, but remain participants of the practice that provides them with their objections"(22). That is to say, critics take their own tradition for granted and deride their opponent for not meeting the standards of such a tradition, which standards now look "objective" only because in using them, the critics forget that they are grounded in the tradition. Thus, for example, .".. in order to find an argument against Aristotle [Popper] would have to find difficulties in Aristotle that are *independent* of the fact that Aristotle does not use the methods of modern science. No such difficulties are ever mentioned. Thus, the 'argument' boils down to: Aristotle is not like us — to hell with Aristotle! Typical critical rationalism!"(63)

Traditions, then, are neither good nor bad; they simply are. They become good or bad (or true or false) only when viewed from the vantage point of another tradition. Relativism raises its flag whenever alternative traditions come onto the field. And in those cases where "objectivity" is demanded (e.g., in value theory), the situation "can be ... rectified by using discoveries ... that correspond to the discovery of alternative geometries"(23). Thus, with dates, coordinates, statements concerning the value of a currency, statements of geometry (after the discovery of Non-Euclidean geometries), we find examples of "many statements that are *formulated* 'objectively' i.e., without reference to traditions or practices but ... still *meant to be* understood in relation to a practice"(23). We should treat normative pretensions such as "theories ought to be falsifiable and contradiction free," in a similar manner. "Continued insistence on the 'objectivity' of value judgments," Feyerabend says, "would be as illiterate as continued insistence on the 'absolute' use of the pair 'up-down' after discovery of the spherical shape of the earth"(23).

Feyerabend knits the fabric of his historical case with yarn from natural philosophy, the theater, and other fields. He most plausibly shows how disputes about standards of rationality turned out to be, time and again, nothing but clashes of different traditions or practices. But is a historical case sufficient? Must we conclude on its basis that reason is practice? Furthermore, where are those discoveries "that correspond to the discovery of alternative geometries?"

The issue, as Feyerabend sees it, is whether we are to think of reason as a disembodied guide of practice or not. Most positions on the problem of reason vs. practice assume that it is, though they disagree on which side should have priority. According to Feyerabend, there are two main such positions: idealism and naturalism. Idealism holds that certain standards should be upheld *come what may*. Idealism is shown to be faulty because the practices that result from upholding the standards are unacceptable to the very proponents of the standards (cf. the required violation of the standards for "progress" to come about in science). In naturalism, "reason receives both its content and its authority from practice. It describes the way in which practice works and formulates its underlying principles"(24). Naturalism is not satisfactory either, even though "having chosen a popular and successful practice the naturalist has the advantage of 'being on the right side' at least for the time being. But a practice may deteriorate; or it may be popular for the wrong reasons ... Basing standards on a practice and leaving it at that may forever perpetuate the shortcomings of this practice"(25). "The inadequacy of standards often becomes clear from the barrenness of the practice they engender, the shortcomings of practices often are very obvious when practices based on different standards flourish"(25).

With this account, Feyerabend takes on both sides of the dispute in contemporary philosophy of science between the logical approach (still the dominant school) and the *historical* one (Kuhn, Polanyi, and, Feyerabend thinks, Lakatos). A much more realistic suggestion "can be illustrated by the relation between a map and the adventures of the person using it ... Originally maps were constructed as images of and guides to reality and so, presumably, was reason. But maps, like reason, contain idealization ... The wanderer uses the map to find his way but he also corrects it as he proceeds, removing old idealizations and introducing new ones. Using the map, no matter what, will soon get him into trouble. But, it is better to have maps than to proceed without them. In the same way ... reason without the guidance of a practice will lead us astray while a practice is vastly improved by the addition of reason"(25). This account is not quite satisfactory because in it "reason and practice are still regarded as entities of different kinds"(26). In his own account, reason and practice are of the same kind, we have relativism, and given the nature of a free society reason (as exemplified by the scientific tradition) is not given any preference.

Are reason and practice of the same kind? Feyerabend's historical case suffers from several defects. The first is that he often confuses the distinction between formal and informal with the distinction between reason and unreason. (25,65) And so when he wants to argue that "reason" is practice, he argues instead that a formal system is still a practice: "But complex and implicit reason is still reason and a practice with simple formal features hovering above a pervasive but unnoticed background of linguistic habits is still a practice"(26). It is true that in many instances, the two distinctions have coincided, but that has not always been the case. Even in his example of Platonists vs. Sophists, each group thought that reason was on its side (even if "rational" is derived from "ratio" — of numbers — "reasonable" is not. And Aristotle, not long afterward, thought it most unreasonable to expect physics to conform to mathematical models. Similar remarks may be made about many of Feyerabend's other illustrations, e.g., the modern dispute between formalists and ordinary language philosophers).

Nevertheless, this is a minor objection, for it does not affect the claim that reason, whether expressed in formal or informal standards, is practice. Of greater significance seems to be a second objection. Feyerabend wants to show that reason is not a "disembodied guide" but a practice. But it seems that for any one tradition or practice (for many of them at least) we could come up with many methodologies or "standards of rationality," and we would be very hard put to say that we had not one, but many practices instead. Significant episodes in the history of science are alternatively described as "inductivist," or "falsificationist," or according to the methodology of "research programs." The point is not that science is not one but many practices, as Feyerabend himself says, but rather that for any one practice within science there are many methodologies already available, let alone possible. Thus, standards of rationality may well be abstractions or disembodied guides of practices.

Even if reason is not practice, Feyerabend's case for relativism is not considerably weakened. Nonetheless, his emphasis is misplaced: What he actually needs to show is that reason is *dependent* upon practice. If so, reason cannot serve as an arbiter of traditions because it is itself dependent on some traditions and not a super-standard hovering over all traditions. By "dependent" I mean that standards of rationality can be seen to be appropriate only because they are connected (in a dialectical fashion, perhaps) to certain traditions, with which they may have arisen. To see this point it will be sufficient to realize that even within science, changes in practice may well lead to changes in standards of rationality, and that different practices (at least all-encompassing scientific practices or traditions, e.g., what Kuhn calls paradigms) may have different standards of rationality.

In arguing for these two last claims, Feyerabend goes beyond his previous case against method. What Feyerabend has attacked all along (since *AM*, that is) is the application of standards come what may, whether such standards are stated flat out (plain idealism) or in conditional form (contextual idealism). But, he is not a naive anarchist who rejects the use of all standards, as can be seen from his attempts to show how some procedures actually helped scientists. In the section "On the Cosmological Criticism of Standards," Feyerabend claims that "the standards we use and the rules we recommend make sense only in a world that has a certain structure. They become inapplicable, or start running idle, in a domain that does not exhibit this structure"(34). The demand for content increase, for example, arose in the first place "from the wish to discover more and more of a nature that seemed to be infinitely rich in extent and quality"(34). The most effective way to criticize any such standard, then, is to cast doubt upon the cosmology within which it makes sense. But this is best done by developing an alternative cosmology (cf. the arguments for proliferation in *AM*). Thus, again for example, the standard of content increase is not untouchable: "It is in trouble the moment we discover that we inhibit a finite world"(34). How could we discover such a thing? By developing theories that describe such a world and which "turn out to be better than their infinitist rivals"(34).

Any one standard, including the one of self-consistency, is up for grabs if we make the appropriate discoveries — i.e. if we produce a successful rival cosmology that has different standards. If our most cherished standards seem impregnable it may only be because the level of criticism is very low, i.e., because the cosmology from which they derive their popularity is not confronted with serious alternatives. It seems to me, that Feyerabend's previous criticism of method by historical examination can be subsumed under this approach. Falsificationism, say, is strongly criticized when we see that adherence to it would have favored the Aristotelians over Galileo, but that Galileo's cosmology, in spite of his counter-inductive procedures, is *deemed* superior. At any rate, the moral of the present line of argument is that standards seem appropriate only because, in the last instance, some theories are thought to be successful, viz. the appropriateness of standards *depends* on their connection with certain practice or practices.[2] Sever the connections (by showing, for example, that the practices do not, and could not, follow the standards) or replace the practices by others that do not follow the standards, and "reason" falls by the wayside. Reason, then, is tied to specific practices or traditions in such a way that it cannot serve as *the* arbiter of traditions. Unless, of course, it can be shown that those specific practices are to be preferred to any others. But surely this can no longer be done merely by pointing out that those practices are in accordance with reason.

Two important objections must be considered at this point. First, standards are to be replaced when the results of research favor a cosmology opposed to that which supports the standards. "But," Feyerabend poses the objection, "how will the results of this research be evaluated if fundamental standards have been removed? ... Does not a decision to accept unusual theories and to reject familiar ones assume standards and is it not clear, therefore, that cosmological investigation cannot try to provide alternatives to all standards?"(37) To be more specific, "when and on what grounds shall we be satisfied that research containing inconsistencies has revealed a fatal shortcoming of the standard of non contradiction?"(37) Some would think that this question would pin Feyerabend on the ropes, but not at all. "The question makes as little sense," he says, "as the question what measuring instruments will help us to explore an as yet unspecified region of the universe. We don't know the region; we cannot say what will work in it"(37). As an example, he suggests that we consider the question of how to measure the temperature of the sun, asked in about 1820. "The very same applies to standards," he says. "Standards are intellectual measuring instruments; they give us readings not of temperature, or of weight, but of the properties of complex sections of the historical process. Are we supposed to know them even before these sections have been presented in detail? Or is it assumed that history, and especially the history of ideas, is more uniform than the material part of the universe? That man is more limited than the rest of nature?"(37) Furthermore, "Standards which are intellectual measuring instruments often have to be *invented*, to make sense of new historical situations just as measuring instruments have constantly to be invented to make sense of new physical situations"(29). Thus, "One may base judgments and actions on standards that cannot be specified in advance but are introduced by the very judgments (actions) they are supposed to guide and one may even act without any standards, simply following some natural inclination"(28).

I find this response appropriate, even though, as I will argue later, it can be turned against Feyerabend's own defense of Aristotle. It seems, then, that science cannot be given preferred treatment by society on the grounds that it (and it alone) follows the dictates of reason (where "reason" is defined in terms of standards of rationality). The first problem is that science neither adheres nor could adhere to the standards so far proposed by philosophers of science. The second, is that even if we found that science accords with reason, reason is dependent upon practice and thus, not a suitable arbiter of traditions. It seems to me, that the sharp pangs brought about by the first problem may be blunted by a different conception of reason, but this matter can be better handled elsewhere.

Now, the relativism so far espoused has to consider the problem of incommensurability (surely, if not only the cosmologies, but even the standards of rationality may differ from tradition to tradition). Two theories are incommensurable if it is impossible to establish the standard deductive relations between them. For incommensurability to occur "the situation must be rigged in such a way that the conditions of concept formation in one theory forbid the formation of basic concepts of the other"(68). This has a most unwelcome consequence for most rationalists in that "we certainly cannot assume that two incommensurable theories deal with one and the same state of affairs (to make the assumption we would have to assume that both at least *refer* to the same objective situation. But how can we assert that 'they both' refer to the same situation when 'they both' never make sense together?)"(70). This is particularly hard to swallow for the many who still feel that realism is at the heart of the scientific enterprise. Unlike the Positivists, the realists believe that the world contains much more than observations; it is the task of science to gradually discover the things that make up the world and to determine their properties and mutual relations, but without changing the things, the properties, or the relations. Realism may be interpreted, Feyerabend says, "as a *particular theory* about the relation between man and the world or as a *presupposition of science*"(70). Those who accept the second alternative are dogmatists. And to show that the first is incorrect, "all we need to do is to point out how often the world changed because of a change in basic theory"(70).

But does the world change when we change theories (or paradigms, or traditions)? Is the matter settled by whether incommensurability is par for the course in theory changing? Feyerabend obviously thinks it is, but in doing so he does not take advantage of the resources relativism has to offer. He should realize that within a relativistic scheme the question about the world changing makes no sense, just as it makes no sense (within the Special Theory of Relativity) to ask whether the mass really changes when we change frames of reference.[3]

This finishes my presentation of Feyerabend's epistemological position in *SFS*. An obvious consequence, he thinks, is that the basic structure of a free society is a *protective structure* which ensures that "*all traditions are given equal rights, equal access to education and other positions of power*"(30). For "if traditions have advantages only from the point of view of other traditions, then choosing one tradition as a basis of a free society is an arbitrary act that can be justified only by resort to power"(30). Thus, epistemological relativism leads to political relativism, the view that all traditions have equal rights.

This view is to be contrasted with the view of today's liberal intellectuals who permit only "*equality of access to one particular tradition* — the scientific, rationalistic tradition of the White Man"(76). This rationalism (science) is made the basis for society, while other views may be *heard* but not permitted "a role in the planning and the completion of fundamental institutions such as law, education, economics,"(144) to say nothing of medicine (175-176). "The excellence of science is *assumed*, it is not *argued for* ... everything else is Pagan nonsense"(73). This assumption of the inherent superiority of the New Church "has moved beyond science and has become an article of faith for almost everyone"(74). "We accept scientific laws and facts, we teach them in our schools, we make them the basis of important political decisions, but without having examined them and without having subjected them to a vote"(74). And when citizens object to the scientists' "working out their private fantasies"(134 at the taxpayers' expense and without public supervision, the matter is presented as if "a peaceful crowd of quiet and self-paid researchers is to be rudely disturbed by an unscientific Gestapo"(212).

Since there is no separation of science and state, science experts are given the power to mold society according to their wishes and consequently "the very same enterprise that once gave man the ideas and the strength to free himself from the fears and the prejudices of a tyrannical religion now turns him into a slave of its interests"(75). In medicine, "the power of the medical profession over every stage of our lives already exceeds the power once wielded by the Church"(74). And in education, "physics, astronomy, history *must* be learned; they cannot be replaced by magic, astrology, or by a study of legends"(74). But in a truly free society, citizens have a say in running the institutions to which they contribute. Experts may be listened to, but their *advice* need not be heeded. After all, "a democracy is an assembly of mature people and not a collection of sheep guided by a small clique of know-it-alls"(87). Maturity is only learned by active *participation* in making decisions; "participation of laymen in fundamental decisions is, therefore, required even *if it should lower the success rate of the decisions*"(87). And since "the only way of arriving at a useful judgment of what is supposed to be the truth, or the correct procedure is to become acquainted with the widest possible range of alternatives,"(86) ... no tradition should be closed to the citizen, who may then "if he falls ill ... be treated in accordance with his wishes, by faith healers, if he believes in the art of faith healing, by scientific doctors, if he has greater confidence in science"(86). Furthermore, if the taxpayers want their state universities to teach Voodoo, folk medicine, astrology, rain dance ceremonies, "then this is what the universities will have to teach"(87).

To those who think that laymen would not have the knowledge necessary to make the required decisions, Feyerabend points out that laymen and dilettantes can, and often do, discover mistakes in the most cherished views of the experts, and that, moreover, expert opinion is often prejudiced, untrustworthy, and in need of outside control.(86-89) Interesting as his arguments are, the rationalist may still reply that in spite of its faults (after all, they do not claim certainty for science; they know it is a self-correcting enterprise) science is a more worthy tradition than magic, astrology and the like. But how can this position be maintained in the face of the epistemological case within Feyerabend's reach? It cannot be shown that any of the standards of rationality are compatible with what rationalists call the success of science. And even then, reason is dependent upon specific traditions and cannot serve as an arbiter of traditions. Having argued that the excellence of science cannot be established on methodological grounds, Feyerabend apparently only needs to argue that such excellence cannot be inferred from the results science produces either. Once these two points are established, the rest are details on the way to replace scientific or philosophical expertise by citizen initiatives and committees of laymen.

I believe, however, that Feyerabend is wrong: the excellence of science can be shown on both counts.

It seems to me that a society has good grounds for "favoring" an ideology over others if there is a communal enterprise for which some ideology or other is required ("ideology" in Feyerabend's sense of "tradition," "practice," and so on). This would no more violate the rights of other traditions than the practice of awarding contracts to carry out certain tasks violates the rights of the individual citizens not chosen. We may grant Feyerabend the point that no one ideology is *inherently* "superior" (or "liberating") without having to conclude that they all are equally well fitted for any task at any given time. Some of them are obviously at a disadvantage (and not just because the dispute has been "rigged" in science's favor), otherwise there would be no need for Feyerabend to argue that they may still make a comeback, as atomism and other views did. Thus, the protective structure of a free society may have to ensure that such views are not stamped out by institutional means, but without having to ensure that they be taught on an equal footing with science. Still, why should science be afforded such preference? We shall see presently.

According to Feyerabend, the excellence of science on methodological grounds is mistakenly assumed. There are no good arguments to show that Aristotle is inferior to modern science, for example. That Aristotelian science does not follow the standard of content increase, as Critical Rationalists are prone to claim, only shows that Aristotle is different, not that he is worse. The difference between these two ideologies, or between Aristotle and whatever is found to actually do justice to modern science,[4] can be traced to an underlying difference of cosmologies. "Aristotle's cosmos is finite, both qualitatively and quantitatively ... it is viewed by an observer who can grasp its basic structure if left in his normal state, and whose capabilities are fixed and also finite. The observer may use mathematics and other conceptual and physical artifices — but these have no ontological implications"(63). On the other hand, "the cosmos of modern science is an infinite world, mathematically structure, comprehended by the mind though not always by the sense ... There is no stable equilibrium between man and the world though there are periods of stasis when the observer can settle ... in a temporary home"(63). Each of the two standards fits a different world; in order to decide between them we must first decide what kind of world we live in.

Now, this view — that the appropriateness of standards depends on the status of the cosmologies associated with them — should receive historical support. Are there any cases in which procedures were vindicated, and perhaps even emerged, because they best fit the "exigencies of research?" (This is not to say that such procedures have been shown to hold, come what may, for the exigencies of research are bound to change, as Feyerabend suggests). There are such cases. Feyerabend himself claims to have discussed several. Of particular importance is the realization that "the idea that information concerning the external world travels undisturbed via the senses into the mind leads to the standard that all knowledge must be checked by observation: theories that agree with observation are preferable to theories that do not"(35). This standard is a cornerstone of Aristotelian methodology, and the corresponding idea part and parcel of Aristotle's cosmos as described above. But this standard, Feyerabend says, "is in need of replacement the moment we discover that sensory information is distorted in many ways. We make the discovery when developing theories that conflict with observation and finding that they are excellent in many other respects (in Chapter 5 to 11 of *AM, I show how Galileo made the discovery*)"(36, my emphasis). This point applies not only to some naive versions of empiricism, but also to the sophisticate Aristotelian empiricism in which "error beclouds and distorts particular perceptions *while leaving the general features of perceptual knowledge untouched*. However great the error, these general features can always be restored and it is from them that we receive information about the world we inhabit"(60). In contrast, modern science, grounded on Galileo's discovery, "postulated just such global distortions"(60).

To summarize: if Feyerabend is correct, standards can be successfully criticized by making certain appropriate cosmological discoveries. But Feyerabend also claims that Galileo made just such discoveries in favor of modern science and against a cornerstone of Aristotelian science. Therefore if Feyerabend is correct, the excellence of modern science over Aristotle has been established *on methodological grounds*.

Next let us consider Feyerabend's argument against the excellence of science on the basis of its results. The rationalist argument is acceptable "only if it can be shown (a) that no other view has ever produced anything comparable and (b) that the results of science are autonomous, they do not owe anything to non-scientific agencies"(100). According to Feyerabend neither assumption survives close scrutiny. Regarding the first assumption, he claims that non-Western traditions did not get a fair fight: they were simply suppressed by the Western colonizers(102). In cases where the oppression has been removed, non-scientific ideologies have shown that they can become powerful rivals of science. For example, when the Party restored free competition between science and traditional Chinese medicine, it was discovered that "traditional medicine has methods of diagnosis and therapy that are superior to those of Western scientific medicine. Similar discoveries were made by those who compared tribal medicines with science"(103). But apart from this one rather special case, what support has Feyerabend for his contention that ancient and "primitive" alternatives "are often more adequate and have better results than their Western competitors and describe phenomena not accessible to an 'objective' laboratory approach?"(104)

Fire. "The inventors of myth invented fire and the means of keeping it." They also invented the rotation of crops, domesticated animals, bred new types of plants, and "crossed the oceans in vessels that were more seaworthy than modern vessels of comparable size and demonstrated a knowledge of navigation and the properties of materials that conflicts with scientific ideas but is, on trial, found to be correct"(104). This is the bulk of the evidence.[5] Now, the issue is not whether ancient man was stupid. Of course he was just as smart as we are. After all, he was our genetic equal.[6] And apparently he had plenty of time to think. So it is not surprising that he was able to do lots of worthwhile things. But should we conclude with Feyerabend that "if science is praised because of its achievements, then myths must be praised a hundred times more fervently because its achievements were incomparably greater?"(104)

Now, which phenomena have been described here that are "not accessible" to science? I find none. But perhaps the results of myth are incomparably greater? (Though comparably would do). I think that Feyerabend is confusing two levels of evaluation. I am sure that many people have wondered about the genius needed to invent fire or breed plants, or to do for the first time any of the other things in the list, practically out of the blue. Considering the situation at the time, those achievements may be indeed more extraordinary than many in our day, given what we already know and the means that are available to us. Greatness often depends on time, place and circumstance. It would be preposterous to compare Galileo to the average graduate student of physics and astronomy today. Galileo is one of the greatest giants in our intellectual history; no one would make similar claims about the other. But when we compare results, the matter is altogether different. The graduate student has at his command a host of intellectual and technological tools that would permit him to do, on the whole, far more and with more precision than Galileo could. Modern physics and astronomy have been immensely refined since Galileo's day. On one level of evaluation, Galileo is a giant, the other a midget. But when it comes to results, the student's science is more excellent by far. Similarly, modern man can have fire in thousands of ways not available to the ancients, and can do with it millions of things unimaginable to the "primitives." As for vessels, one can not help but picture an ancient seaman's reaction to speed boats, luxury liners, aircraft carriers and atomic submarines, to say nothing of our travels through the ocean of air. It is remarkable that they could navigate so well in those days. But today we can do better, not only in and under water, but as far away as Jupiter and Saturn. And when it comes to manipulating the living world, Feyerabend seems to forget all about the awesome powers that molecular genetics, to name only one field, has begun to unleash in recent years. I could go on, but I think it is clear by now that the comparative excellence of the results of myth that Feyerabend found, stems from his confusing two levels of evaluation. We may be standing on the shoulders of giants, but then even midgets can reach higher and see farther than the giants could.

Even here I am granting too much. The results of modern science have been obtained in a very short time, whereas for all we know it took the myth makers thousands of years and much blundering. More can be said about the depth and breadth of the results of science in other respects, but it is time to move on to the autonomy assumption. "There is not a single important scientific idea that was not stolen from elsewhere," says Feyerabend.(105) Copernicus, for example, got his ideas from ancient authorities. Feyerabend exaggerates — from which ancient myth did Watson and Crick steal the idea for the structure of DNA? — but we may grant him that many important scientific ideas had their origins elsewhere. So what? The fact that Galileo was very instrumental in bringing about modern science does not change the balance of excellence of *results* discussed above. Feyerabend's case is much weaker in the many instances where we could at most speak of inspiration.

When Feyerabend challenges science in its own turf, his comparisons fail. But in other places he favors the view that the excellence of results is in the eye of the beholder. I suppose that may be fair enough. Through navel contemplation some people feel at one with the universe. By disciplined malnutrition, or by the use of peculiar mushrooms or the sniffing of natural glues, some others may have even "traveled the celestial spheres to God himself whom they viewed in all his splendor receiving strength for continuing their lives and enlightenment for themselves and their fellow men"(31). Feyerabend and his mystics are much more impressed by such travels than by moonshots, the double helix, and non-equilibrium thermodynamics. To each his own. But why then, he wants to know, should a free society make science its basic ideology?

Apart from the reasons already given, there are two major points still to be considered. First, scientific ideologies either provide instruments that help us deal with the world in certain ways, or they may perhaps be conceived as being such instruments themselves. If those ways of life are agreeable to a society, if the communal enterprises to which they lead are likely to confront every citizen, it seems that the society may be not only entitled but even obliged to inform its future citizens about them, i.e. to give its citizens knowledge of the relevant scientific ideologies. For example, if for all a society knows, its survival depends on a certain agricultural technology, it is extremely reasonable for that society to educate its children on the basis of that technology. As it happens, our modern societies depend on many diverse applications of science. We find ourselves in the midst of cars, airplanes, electric appliances and power plants; we heat and cool our houses and harvest and process our foods with technological means. This may not be the best way to live. But it is the way we live, and our future citizens need a minimum of information about the sort of world that will confront them. For a society that inhabits a region full of lakes and streams, knowledge of swimming and fishing may be important for survival. In our societies, some elementary acquaintance with the principles of electricity and dynamics may be necessary to ensure the preservation of life and limb. And better than a passing acquaintance is often needed to manage more than mere survival. So future citizens have to learn about science.

Of course, a free society can only educate its future citizens as to such instruments, but it should not force them to lead their lives accordingly. And if the citizenry decides to concentrate on navel contemplation from now on, science may be useless to it, and so there would be no point in giving science a fundamental role in education. But as long as our world decides to live by the fruits of science, the state would be remiss unless it makes the scientific tradition a cornerstone of education.[7]

Feyerabend was correct in pointing out that to suppose that the very character of a free society is inextricably bound with the ideology of science is to assume the inherent superior value of science. He was also correct in claiming that the rationalists had no good arguments for such assumption. But his own arguments against the excellence of science failed, making possible a case for the prominent position of science on grounds of methodology and of results. That prominence is not inherent, however, but relative to the exigencies of research in our historical period and to the wants of modern societies (even if both standards and wants arose together with modern science). Just as a society, by choosing the contractor that best fits its present needs, does not trample on the rights of all other bidders, society is also entitled to favor the scientific ideology that offers the most desirable way of life. But none of the foregoing goes to show that a free society ought to choose the way of science.

Nevertheless, it seems to me that even this last hurdle can be overcome. "What is so great about science?" asks Feyerabend, "And what is so great about knowledge?" Let us turn the tables on him and ask: "What is so great about a free society?" It takes more than majority rule for a society to be free; it takes a commitment to allow others an opportunity express unusual points of view and to pursue lifestyles of their own choosing. But why should the majority permit this? After all, it may find those other opinions heretical and those other life styles an affront to the sensitivities of the many decent people who have the power.

That the majority should is best explained by J.S. Mill, whose arguments, according to Feyerabend, cannot be improved.(86) In *On Liberty*, Mill argues that man's knowledge is not infallible, and that as a consequence, no opinion should be denied a hearing by the majority of citizens, for "if the opinion is right, they are deprived of the opportunity of exchanging error for truth; if wrong, they lose, what is almost as great a benefit, the clearer perception and livelier impression of truth produced by its collision with error"(*OL* 21).[8] In actual practice, though, no view is completely right or wrong, and so progress "for the most part only substitutes one partial and incomplete truth for another; improvement consisting chiefly in this, that the new fragment of truth is more wanted, more adapted to the needs of the time than that which it replaces"(*OL* 56)(cf. the dependence of standards on the exigencies of research). A free society is justified, then, because "the only unfailing and permanent source of improvement is liberty, since by it there are as many possible independent centers of improvement as there are individuals"(*OL* 86). Thus the pursuit of knowledge is a most crucial element in the justification that Mill offers for a free society; that pursuit, moreover, requires proliferation. It is clear also that when it comes to knowledge — of the nature of the world and of man's place in it — cosmological knowledge is of the greatest importance. This cosmological knowledge will not be any one particular ontology but rather an enterprise that lives by proliferation and changes ontologies and standards so as to best suit the needs of the time. But what is this enterprise if not science as described by Feyerabend (or at least science as Feyerabend would ask for in his letter to Santa Claus)? "Science is essentially an anarchistic enterprise," said Feyerabend in *AM*(p.17), and thereby shocked many. But he was actually extending Mill's lessons to the epistemology of science.[9] In the first part of *SFS*, these lessons are further strengthened by Feyerabend's case for epistemological relativism. Therefore, science properly understood, is closely interwoven with the fabric of a free society and to a large extent justifies it.

For this, and for the other reasons discussed above, Feyerabend's thesis of the separation of science and (free) state does not succeed.

Two other matters are still in want of further comment, however. Even if the prominence of science can be justified, that prominence is limited and should not result in the obliteration of other views by institutional means. In education, for example, Feyerabend makes us realize that too much emphasis is placed on the acquisition of information and not enough on the preparation to do the independent and mature thinking essential to the full members of a free society. To challenge established points of view and to develop and consider alternatives should be far more prevalent than it is now. But certain things students ought to be acquainted with, and so the study of physics, astronomy, and history should not be replaced by that of magic, astrology, and legends. Feyerabend's arguments for proliferation show that other points of view should be not only permitted but encouraged. They do not show, however, that those alternative points of view should be given as fundamental a role as that of the scientific ideology so far best adapted to our historical circumstance.[10]

If education is a matter in which Feyerabend goes too far, medicine is one in which Feyerabend has excelled in the art of diatribe and rationalist baiting. The lessons that Feyerabend draws for science from his examination of "scientific" medicine are blunted considerably because it is not clear just how "scientific" that medicine is. Cancer research, for example, would fit neither Kuhn's nor Lakatos' criteria for mature science. Medicine is actually a mismatch of things: some applications of modern biology, some dressing up of the practices of the medieval hypothecaries, and not a clear vision anywhere. So it is not surprising that contemporary medicine is good at some things and terrible at others. With the advent of molecular genetics and other powerful theoretical tools in biology, it is to be hoped that "scientific" medicine will truly become applied biological science. To integrate the discipline this way, all kinds of sources will have to be tapped. Nothing short of arrogance can then justify the purge of alternative ways of healing. It is not surprising that ancient or "primitive" medicines are better in some areas than "scientific" medicine. But to acknowledge this point is not to give support for specific claims. Here, Feyerabend is most disappointing. How does he know, for example, that women with breast cancer are better off going to acupuncturists, faith healers, and herbalists? Because he advised some of them and followed the fate of others.(206) I think that Feyerabend has made such a big deal of the shortcomings of medicine that he owes his beleaguered rationalists better than anecdotal accounts of this sort.[11] Who knows? A better account may be a public service.

This is not to say that he has not already provided a service by unmasking bad arguments, unwarranted assumptions, and intellectual complacency. As usual, he has gone too far. But philosophy of science can well afford bold thinkers who are prepared to defend implausible ideas against all corners.

Notes

1. *Science in a Free Society*, NLB (London: 1978), page numbers will be given in parentheses.
2. Feyerabend argues that the meaning of standards also depends on such connections.
3. For a more detailed account of relativism and the required "discoveries," see Ch. 2 of my *Radical Knowledge* (to which Feyerabend also refers his readers).
4. The following, then, should not be taken as a defense of the condition of content increase, which is besieged by many other problems (cf. *Am*, Ch. 15).
5. Other items in Feyerabend's list are very questionable. For example, the myth makers "developed an art that can compare with the best creations of Western man." Even if true, how is this relevant to the issue? "The best ecological philosophy is found in the Stone Age." The best? How? As for his remarks on medicine, see mine on pp. 26.
6. Although some of Piaget's researches indicate that "primitives" do not reach some of the higher intellectual stages available to all humans. See his *Psychology and Epistemology*, The Viking Press (New York: 1972), pp. 61-62.

7. And if the society wants to maintain and improve its technology on a grand scale, it will have to provide adequate support for sceintific research in general, since there is no telling in advance where exactly technological opportunities will be born. This reason is implicit in many appeals for funding of scientific programs, not the notion that science is inherently superior. It does not help to claim, as Feyerabend does, that we will find plenty of "willing slaves" to pick up the scientific slack. A society not committed to supporting science does not offer many of its citizens the opportunity to become scientists.
8. This and the following page references are to the Bobbs-Merrill edition (New York: 1956).
9. Feyerabend no longer urges everyone to "think, feel, live through a competition of alternatives"(144). "I do not show that proliferation *should be used*, I only show that the rationalist cannot *exclude* it"(145). And this point is made via "an argument that derives proliferation from the monist's own ideology"(145). Of course, one can hold that proliferation is essential not only for progress in science but for the fabric of a free society, without being a "mealymouthed wishywashy nobody who understands anything and forgives everything"(148). One may have very strong and dogmatic beliefs and one may also feel contempt for others views, while only being required "to let them have their say and not to stop them by institutiontal means."
10. Most people, as Mill would recognize, think of education only in terms of information, and would be outraged if Voodoo becamse part of the curriculum of universities. As an example, Feyerabend may consider the peasants' hostile reaction to changes in the educational system of his native Austria in the earlier part of the century. Wittgenstein, who served as school teacher in the movement to enable the peasants to become independent and mature thinkers, was a victim of the reation. Similarly, the most political intervention is likely to suppress proliferation and not to encourage it.
11. Feyerabend does give some references. He mentions I. Illich's *Medical Nemesis* and Coulter's *Divided Legacy*. (In subsequent work his case against Western medicine improved considerably.)

Appendix B
A *Rehabilitation* of Paul Feyerabend

Introduction

One of Paul Feyerabend's pet peeves was the way the history of philosophy had maligned Ernst Mach by treating him as the founding father of logical positivism. Feyerabend often spoke of writing a *rehabilitation* of Mach and did take a few beginning steps in that direction.[1] I largely agree with Feyerabend's appraisal of the historical misjudgement of Mach, but my purpose in this paper is not to explore that issue: my intent is rather to write about Feyerabend as he would have about Mach. For it seems to me that few philosophers have been as misunderstood as Feyerabend himself.

I am particularly interested in discussing the following:

(1) Feyerabend's alleged claim that "anything goes";
(2) his alleged "incommensurability" thesis;
(3) his alleged relativism.

Before discussing Feyerabend's views, however, I would like to say that over his career Feyerabend changed his mind about a variety of subjects, sometimes drastically. He had a right to grow intellectualy — he liked to remark — to try many approaches. That is the least we should have expected from such an able proponent of pluralism. This should be remembered as I give a sketch of Feyerabend's epistemology of science circa the publication of the first edition of *Against Method (1975)*. As I go along I will try to point out those instances where Feyerabend changed his mind significantly.

One last preliminary point: What I am about to do could be seen as offering a *theory* about Feyerabend's philosophy: a systematic and consistent account of it. This he would have hated very much. He disliked theories of knowledge because he thought that knowledge, a complex part of human life, was always changing, as it should. He swore off relativism because relativism was a *theory* of knowledge. It would thus not have pleased him to think that his "philosophical life" could be so easily "nailed down" in just one paper, that a "neat" account of it could so easily "freeze" what should have been fluid and open-ended.

Now, Feyerabend's position on the three issues I wish to discuss is grounded on his analysis of the history of science. It is from his understanding of the practice of science that much of his philosophy follows in a principled way, and it is thus to that understanding that I will first devote my attention.

1. Feyerabend's Analysis of the History of Science

It was Feyerabend's philosophical analysis of crucial episodes in the history of science that first led him to suggest his famous (or infamous) "Anything goes," his notion of incommensurability, and his epistemological relativity.[2] Of those crucial episodes, Feyerabend's favorite was Galileo's rebuttal of the formidable objections from physics, astronomy, philosophy, and religion against the motion of the Earth. Empiricist history of science had often painted Galileo as a patron saint of its main credo: That experience is the final judge of theory. There are disputes among the various forms of empiricism as to whether experience (the "facts") can justify a scientific theory, but in a crucial way most, if not all, agree that a theory in conflict with experience is not acceptable. Ironically, Galileo's own works gave Feyerabend his most powerful argument against the empiricist credo.

Among the many telling objections against the motion of the Earth, perhaps the Tower Argument presented the strongest of all. It goes as follows. Suppose that you let go of a stone from the top of a tall tower. If the world moves, by the time the stone hits the ground, the tower being stuck in the Earth, will have moved considerably (the velocity of rotation of the Earth would have been calculated to be about 1,000 miles per hour). Thus there should be a perceptible difference between the initial and final distances from stone to tower. But when we actually look, there is no difference at all! For the distance to remain constant, if the earth did move, the stone would have to fall in a curved path, but we clearly see the stone fall straight down. Therefore the Earth does not move.

Faced with the Tower Argument what could Galileo say in defense of the Copernican view? He said that the stone does not fall straight down, no matter how clearly we see it. The stone only *seems* to move so; its *real motion* is far more complicated than that. But this makes no sense, people thought: Motion is *observed* motion. Not so, said Galileo. Shared motion goes unobserved (motion is relative).

The reason why the stone keeps its distance from the tower is that its real motion has two components: the first is straight down and we notice it, the second (which Feyerabend called circular inertia) is shared with the

Earth, the tower, and the observer (us). That is why we do not notice it; but it is there all the same. Just as the tower moved laterally, so did the stone. When we are in an airplane flying smoothly we do not *perceive* that the passenger sitting next to us and the magazine on our lap are traveling at 650 miles an hour, even if we know they both are. We do perceive the motion of the stewardess up and down the aisle and of the drink spilling on the sleepy man to our left. We perceive those motions that we do not share, but fail to notice those that we do. Galileo made the same point by means of examples about ships, and in this manner he neutralized the objection against the Copernican view.

What conclusions did Feyerabend draw about the Tower Argument then? People noticed a phenomenon and *interpreted* it in what they thought was the most *natural* way, i.e., the stone *moves* only straight down. It was this *natural interpretation* of the phenomenon, not the phenomenon itself, that contradicted the Copernican view. Galileo did away with the contradiction by providing a *different set of natural interpretations*. Galileo, then, constructed a new empirical basis! This new empirical basis, furthermore, is constituted by a *new theory of interpretation* congenial to Copernicus.[3]

What permitted the change of empirical basis was, in part, a change in theoretical assumptions. More specifically, Galileo changed the concept of motion. Motion had been supposed by his opponents, the Aristotelians, to be only observed motion. In the jargon of this century, we might say that the Aristotelians had an operationalist concept of motion (i.e., that a phenomenon would count as motion only if it could be expressed in terms of observable changes). But Galileo introduced into motion components that could not be observed. And in the particular case of the Tower Argument, one of those components was circular inertia, which in addition to being highly theoretical (and *ad hoc*) was quickly abandoned after the Copernicans won the scientific revolution.

Philosophers have often praised Galileo for preferring his eyes to his Aristotle,[4] but Feyerabend's analysis shows how misguided that praise has been. Galileo was a Copernican. The central thesis of Copernicus stated that the Earth was one more planet in orbit around the sun. The evidence of the eyes, unfortunately, refuted this Copernican thesis. As we well know now, the Earth and Venus are sometimes on the same side of the sun, and thus rather close to each other; and sometimes they are on opposite side of the sun and thus very far from each other. The same can be said for Mars, except that Mars gets much farther from the Earth. It seems commonsensical then that Venus and Mars should look brighter when they are close and dimmer when they are far. But the magnitude of Venus barely changes, and that of Mars does not change as much as it should.

Nevertheless Galileo's admiration for Copernicus did not decrease, even though Copernicus, in Galileo's words, "with reason as his guide... resolutely continued to affirm what sensible experience seemed to contradict."[5] Reason, it appears, can overturn the verdict of experience: "there is no limit to my astonishment," Galileo writes, "when I reflect that Aristarchus and Copernicus were able to make reason so conquer sense that, in defiance of the latter, the former became mistres of their belief."[6] In his own case, Galileo says, he came upon "the existence of a superior and better sense than natural and common sense," the telescope, which then joins "forces with reason."[7] Fortunately enough, his telescopic observations of Mars gave magnitudes much more in agreement with Copernicus' thesis. And the telescope showed that Venus has phases. When Venus is farthest from us, we see its full face illuminated by the sun. When it is closest, most of the face that it shows to us is dark. Thus the amount of light that reaches us from Venus remains constant enough for our eye to perceive little change in magnitude.

In this situation we find an outright conflict between natural and artificial sense. Galileo surely saw it that way. He resolved the conflict by *denying the testimony of his eyes* and siding with the sense that agreed with Copernicus, the telescope. This conflict between senses should dispel any illusions to the effect that scientific instruments merely amplify and sharpen our senses. But some empiricists may feel relieved that it was telescopic *experience* after all that decided the matter. Unfortunately, such relief is unjustified unless they can show that the telescopic experience was without question superior to that of the eye, as far as the observers of that day were concerned, *and* that the choice of the telescope over the eye does not require theoretical assumptions.

Empiricists will have difficulties on both counts. Keppler, for example, wrote to Galileo

> I do not want to hide it from you that quite a few Italians have sent letters to Prague asserting that they could not see those stars (the moons of Jupiter) with your own telescope. I ask myself how it can be that so many deny the phenomenon, including those who use a telescope. Now, if I consider what occasionally happens to me, then I do not at all regard it as impossible that a single person may see what thousands are unable to see....[8]

Feyerabend had many interesting things to say about the Italian observers' troubles with Galileo's telescope, but for the sake of brevity let me discuss only the second difficulty.

The second, and major, difficulty for empiricism is that Galileo's trust in the telescope required the granting of several *theoretical* assumptions. Images from the heavens would travel immense distances, enter a different medium upon hitting the Earth's atmosphere,[9] work their way through the telescope, and finally be handled by a brain that had never perceived anything like them. To be assured that those images were not significantly distorted, Galileo needed supporting theories about optics, about the nature of light, about the atmosphere, about the interaction between light and a variety of gases, about the telescope, and about perception. We may realize, then, that it was not Galileo's telescopic *observations* that challenged the geocentric view of the universe, but Galileo's observations together with a host of assumptions from many supporting or auxiliary sciences. The crucial question was: Could experience *alone* have reconciled the magnitudes of the planets with Copernicus' thesis? If by experience we mean sensory experience, the answer is no. If we allow telescopic experience, then we should remember that such experience could be taken as reliable only if interpreted on the basis of certain theories. The answer again is no.

To make matters worse, most of the auxiliary sciences in question were not within Galileo's reach. Some of them took hundreds of years of development before they could fully back Galileo's hunches. Thus to a good empiricist of the day, many of Galileo's theoretical assumptions should have seemed unwarranted.

Galileo's scientific instincts may seem unerring to us, with the benefit of hindsight, but that is a different issue. Nevertheless, let us turn the tables and see why we should recommend to Galileo to act as he did. The same analysis that leads to the realization that Galileo made theoretical assumptions applies to the eye as much as it does to the telescope. Visual perception is a complex process in which the brain takes into account not just the "input" from the retina but also from the inner ear and hundreds of skeletal muscles (to determine the position of the body) and from the other senses, as well as from memory and imagination. Think of how vague images suddenly come into focus when we smell the particular scent of a flower in a forest or hear the growling of a dog in a dark street.

The working of perception in these and other complex ways is the result of the history of adaptations of the brains of our ancestors to a variety of environments. The extent to which the senses can be "trusted" is thus not a matter for philosophy alone to determine. Psychology tells us of the richness and complexity of perception, neuroscience may help reveal the structures that make such richness and complexity possible, and evolutionary biology may explain how those structures arose and may also give us clues about where they apply.

Trying to decide between the eye and the telescope is a very complicated affair that involves a host of theoretical assumptions. Of course, Galileo did not know anything about neuroscience, let alone evolutionary biology. But he did realize that his opponents' assumption about the reliability of the senses had to be backed up by some view of the relation between the world and the senses. Indeed that view was Aristotle's theory of perception, according to which an undisturbed mind will take on the "form" of an object as long as that form has traveled through an undisturbed medium. That is how a normal observer under normal conditions would gain knowledge of the world. Thus behind the clash of the senses there was a clash of theories, whether those theories were explicit or not. Galileo chose the theoretical direction that promised him the most exciting discoveries.

It is important to notice that Galileo did not merely take methodological shortcuts. It is not as if his hunches had led him more quickly to results that more patient methodologists would have achieved eventually. Not at all. If method requires the priority of experience, method would have forever closed the door on a view that could not be established without overthrowing accepted experience. If, in pursuing a theory that had been refuted by experience, Galileo committed a sin against science and philosophy, we should love not only the sinner but the sin.

This is not to say, however, that theory is always overturning the verdict of experience. Nevertheless, it is to say that it *may*. Thus we should not be startled upon hearing that some brilliant scientist showed little concern about the results of apparently crucial experiments or observations. When Einstein was asked what he would have thought if an important experiment had disconfirmed his theory of general relativity, he answered "Then I would have to be sorry for dear God. The theory is correct."[10]

I trust this sketch suffices to show how Feyerabend's analysis of the history of science led him to uncover some very important limitations of the empiricist credo. Of course, empiricists did not take kindly to the use that Feyerabend, and Kuhn, made of history. Notice, however, that they could not dismiss history (i.e., practice) on the hitherto successful grounds that the epistemology of science deals not with how science is done but with how it ought to be done. For what history showed us, through Kuhn and Feyerabend, was that in order to achieve scientific success, as proclaimed by empiricists, scientists sometimes had to violate the methodological prescriptions of those very empiricists.

An important empiricist counter-attack concerns the nature of the evidence that Kuhn and Feyerabend have used against the standard views in philosophy of science. Kuhn and Feyerabend have used history to make their points. But how do they know that history can be trusted to that extent? How is it that the "facts" of physics, astronomy and chemistry can be overthrown but we must show reverence for the "facts" of history? Kuhn and Feyerabend cannot have it both ways. Indeed, it seems reasonable to suppose that physics is more reliable than history. Thus if Kuhn and Feyerabend's thesis against the theory-observation distinction is granted, their evidence is not worth all that much.

Several important points must be considered in reply. The first is that when the objection is aimed against Feyerabend it fails to take into account the structure of his argument. To see this more clearly it pays to consider first a similar objection. According to this other objection, Feyerabend argues against reason in science. But in order to establish his conclusion, Feyerabend has to use reason. If reason is no good, Feyerabend's means for establishing his conclusion are no good either. Thus Feyerabend must be committed to the correctness of reason. Unfortunately for him, this consequence invalidates his conclusion.

What this objection misses completely is that Feyerabend does not need to be committed to the correctness of reason, whether as rational argument or methodological rules. His argument is a *reductio ad absurdum*, and a rather simple one at that. In a *reductio* one assumes for the sake of argument the opponent's position and then derives a conclusion unacceptable to that opponent (that is, reduces his position to absurdity). Feyerabend drives this point home time and again. He says, for example

> Always remember that the demonstrations and the rhetorics used do not express any "deep convictions" of mine. They merely show how easy it is to lead people by the nose in a rational way. An Anarchist is like an undercover agent who plays the game of Reason in order to undercut the authority of Reason.[11]

Of course, if reason is no good Feyerabend has no argument. But he does not need one, since the conclusion is already established.

Now we can consider the initial objection. How can Feyerabend trust the "facts" of history after undermining the apparently far more solid "facts" of physics, astronomy, chemistry, and the like? Furthermore, is Feyerabend now arguing that methodologies are refuted by facts and thus have to go? But did he not argue against this very way of dispensing with views? (cf. his arguments against falsificationism). Once again Feyerabend takes the cue from his opponents:

> *They* prefer Galileo to Aristotle. *They* say that the transition Aristotle-Galileo is a step in the right direction. I only add that this step not only *was not achieved*, but *could not have been achieved* with the methods favoured by them. But does not this argument involve highly complex statements concerning facts, tendencies, physical and historical possibilities? Of course it does, but note that I am not committed to asserting their truth.[12]

Once again, Feyerabend's argument is a *reductio*. His aim is not to establish the truth of propositions but to make his opponent change his mind (or at least give him pause to consider why he should not change it). To achieve this end, he says

> I provide him with statements such as 'no single theory ever agrees with all the known facts in its domain'. I use such statements because I assume that being a rationalist he will be affected by them in a predictable way. He will compare them with what he regards as relevant evidence, for example, he will look up records of experiments. This activity, combined with his rationalistic ideology will in the end cause him 'to accept them as true' (this is how he will describe the matter) and so he will perceive a difficulty for some of his favourite methodologies.[13]

The assumptions that Feyerabend makes about the efficacy of his rhetorical devices, as well as his motivation, should be of no concern to the rationalist.

> All he *needs* to consider, all he is *permitted* to consider is how the statements surrounding the case studies in my book are related to each other and to the historical material and whether they can be read as an argument in *his* sense. I admit that my procedure succeeds by manipulating the rationalist but note that I manipulate him in a way in which he wants to be manipulated and constantly manipulates *others*: I provide him with material which interpreted in accordance with the rationalistic code creates difficulties for views he holds. Do *I* have to interpret the material as he does? Do *I* have to 'take it seriously'? Certainly not, for the motivation behind an argument does not affect its rationality and is therefore not subjected to any restriction.[14]

It is not clear, by the way, that Kuhn also intends to offer a *reductio*. But the objection does not work against him either. In the first place, Kuhn does not claim that conflict with historical facts makes it obligatory to abandon traditional epistemology of science.

> [The contrary historical facts] by themselves. . .cannot and will not falsify that philosophical theory, for its defenders will do what we have already seen scientists doing when confronted by anomaly. They will devise numerous articulations and *ad hoc* modifications of their theory in order to eliminate any apparent conflict.[15]

Such historical anomalies can then at best "help to create a crisis or, more accurately, to reinforce one that is already very much in existence."[16] Their main contribution to the epistemology of science is that they make possible "the emergence of a new and different analysis of science within which they are no longer a source of trouble."[17]

Apart from the analysis of the reasoning involved, a key point at issue is the overthrow of the empirical basis. If it can be done for physics it can surely be done for history as well. But then, views of science that take such historical facts seriously can have the rug pulled from under their feet no less than theories in physics or astronomy. Nevertheless we do not have much of an objection. After all, the task at hand would require the overthrow of the present socio-historical empirical basis that has been used by Kuhn and Feyerabend. Theories of science are thus in the same boat with scientific theories: they can be replaced by alternatives that change the "facts" on them. But that possibility does not diminish their worth. There will be no decisive objection along the present lines, therefore, unless the empirical basis of this view of science is actually overthrown.

If Feyerabend is right, all methodological principles have exceptions (progress requires their occasional violation). This result alone has made many rationalists conclude that science would then be ruled by anarchy. Feyerabend seems to accept this interpretation when he proclaims that "science is essentially and anarchistic enterprise," but his emphasis is not that science falls short of what it ought to be, but rather that what it "ought to be" is nothing but an unwarranted imposition by philosophers on an enterprise that investigates a complex medium. As he puts it, "A complex medium containing surprising and unforeseen developments demands complex procedures and defies analysis on the basis of rules which have been set up in advance and without regard to the ever-changing conditions of history."[18] A methodologically correct science would be an inferior science.

2. "Anything Goes"

When Feyerabend claims that "the only principle that does not inhibit progress is: anything goes," he is gunning for the philosophy of science. As he says, "*anarchism,* while perhaps not the most attractive political philosophy, is certainly excellent medicine for *epistemology* and for the *philosophy of science.*"[19] He arrives at this result on the basis of his examination of historical episodes and of his analysis of the relation between idea and action. This is by no means a negative, skeptical philosophy, as we will see in this section. "Theoretical anarchism," he says, "is more humanitarian and more likely to encourage progress than its law-and-order alternatives."[20]

Feyerabend's aim is to promote the advantages of pluralism in science, a course of thought that conflicts with Kuhn's ideas, for Kuhn argued that the tenacious commitment to one point of view is the best way to ensure progress in science. An examination of this difference with Kuhn will shed light on some important aspects of Feyerabend's philosophy.

Now, Kuhn correctly points out that a comprehensive view is abandoned not because it has anomalies, but because it is replaced by an alternative. Anomalies thus do not refute a paradigm, but they may bring a crisis about if they are thought to be important enough (for then the failure to assimilate them gains great significance). No anomaly, however, is as important as one which a competitor claims to have explained — no anomaly, that is, accentuates more the loss of confidence in the paradigm. The reason is that, as Kuhn explains, a paradigm is accepted on the promise of future performance, on the promise, that is, that it will prove the best way to conceive of the world.[21] When a competing would-be paradigm seems to be doing better, our faith in the *promise* of our anomaly-besieged paradigm may well falter. Thus, Feyerabend thought, we will create more crises, and therefore more fruitful change in Kuhn's own terms, by providing a mechanism to strengthen the anomalies. To accomplish this goal, science should be organized so as to require the *continuous generation of alternatives*. This Feyerabend calls the *principle of proliferation*.[22]

Expectation can cut two ways: it permits us to see, but it may also keep us from seeing. Can we do better? Yes, if we have more than one set of expectations to draw from. That is Feyerabend's point. Of course, it may be difficult for the same individual to hold alternative sets of expectations. On the other hand, in a discipline where proliferation is not ruled out, some individuals may develop a different world view, and may thus be in a position to point out areas of difficulty that the other members of the discipline may tend to overlook. That is the first step. The second step is that where one side may face contrary facts or fuzzy pictures, the rival may claim consistency with observation (within its system) or clear pictures. At that stage it may be far more difficult for the practitioners of the standard view to regard the anomalies as mere oddities that may be taken care of at a later date. The failure to articulate the paradigm may then acquire great significance precisely because of the pressing possibility that the paradigm's categories may not provide the best box into which to file the experience in question. And this suspicion of doom grows larger in proportion to the ease demonstrated by the rival view.

Two resolutions are then made available: (1) the defenders of the standard view do manage to assimilate the anomaly, or (2) a crisis ensues that may result in a paradigm shift. Either resolution is prompted by the presence of a rival. Without the competition by that rival there would have been no motivation for worry in the first place. That is, even according to Kuhn's own specifications, the scientific discipline would not have done as well. The point is not really all that intricate. It is possible to discover one's own faults through introspection. But the great majority of us learn of our faults thanks to all those around us who are plain delighted to point them out.

On the other hand, from Kuhn we have learned — against falsificationism — that a view should not be dropped merely because some of its predictions fail. A view should be given time to develop, to bear out its original promise. Thus commitment to a view is crucial to science

(Feyerabend calls this the *principle of tenacity*). Where Kuhn goes too far is in his insistence that the *entire discipline* be dedicated to only one view. It is more fruitful for science, Feyerabend says, to have several competing groups working on those acts of ideas that they find particularly promising. It is also a more humanitarian outlook, for instead of authoritarian enforcement of a dominant view, it allows for the individual scientist's pursuit of happiness through his scientific work.

Feyerabend makes use of two simple notions. First, people work best at those things they like best. Thus scientists should be encouraged to develop those views of nature that for any reason have caught their fancy. Second, the quality of work improves when strong challenge points the way in the direction that requires improvement. Without such challenge we become complacent, and even if we want to keep on our toes we may fail to see the flaws in our pet theories because we are too close to them, or because we lack enough imagination, or for many other reasons. Thus competition aides us in our pursuit of excellence.

The principle of proliferation, and within its action, the principle of tenacity, leads to greater human happiness. Those two principles also create the conditions for fruitful change and improvement. Thus both humanity and science are the better for their presence. Kuhn's insistence on the restriction of the principle of tenacity to only one view is the sort of dogmatism that gets in the way of what Kuhn himself considers advantageous. On the other hand, the falsificationist's rigid quest for rejection robs science of the depth tenacity offers to it — without contributing to individual happiness, for scientists are obliged to give up what may well intrigue them. It seems that Feyerabend is offering us the opportunity to have our scientific cake and eat it too.

Feyerabend does not claim that science in fact progresses. He leaves the evaluation of progress up to each individual (or each individual group). He only tells the different parties how they can best bring about their own goals by adhering to the principle of tenacity and permitting the functioning of the principle of proliferation. If by continuous confrontation with competing views a scientist is spurred to improve his view — where improvement is measured by his own standards — then his work has profited. If, on the other hand, he finds that the competition has gotten the best of him — again, for whatever reasons — he abandons his view and takes up the other. In this case he has also profited. The appropriate mix of the principles of tenacity and proliferation increases the scientist's chance of scientific profit *as seen from his own methodological point of view*.

This is just one variation on Feyerabend's grand theme of scientific anarchy. His tune against Kuhn's dogmatism is an adaptation of a general approach that he developed against the standard methodologies. As we saw earlier, Feyerabend shows that even the most obvious methodologies have limitations. For empiricism it makes no sense to use hypotheses that, say, contradict well-confirmed theories, and even less sense to use hypotheses that contradict well-established experimental results. But Feyerabend argues that such hypotheses may be used to the great advantage of science. Can we really advance science by proceeding counterinductively?

Yes. The reason, as we may have surmised from Feyerabend's case against Kuhn's dogmatism, is that those counterinductive hypotheses give us evidence that cannot be obtained in any other way. Prejudice is often discovered not by analysis but by contrast. If, as we have seen, every fact is already viewed in a certain way, and to progress often requires viewing facts in a different way, then we simply need alternative ways of seeing. As for the conflict between those counterinductive hypotheses and the facts — and it is that conflict that presumably makes them counterinductive — we should remember that no theory ever agrees with all the facts in its domain (cf. Kuhn's account of how a paradigm is a promise of results and not a collection of them). If such conflict is grounds for throwing out a theory, then we should throw out all theories. The main reason for not trembling in the shadow of the facts is that facts are constituted, in part, by older ideologies, and thus a clash between facts and theories may actually be an indication of progress — an indication that our probe is coming into contact with some of the principles assumed in familiar observational notions.

It is often said that we cannot step outside science to see whether it represents the world. This simple point is supposed to dog the idea that truth is correspondence to reality. And maybe it does. But we may still observe the relationship between our science and the world by comparing our science with an alternative interpretation of what the world is like. As Feyerabend says, "*we need a dream-world in order to discover the features of the real world we think we inhabit* (and which may actually be just another dream-world)."[23] In this Feyerabend echoes John Stuart Mill. If our present views are right, by criticizing them from another vantage point, we come to understand them better. And if they are not right, we gain the opportunity to replace them.

If this is so, however, we come to realize that any idea, no matter how ancient or absurd, is capable of improving our knowledge. This sounds at first preposterous. For example, we finally got rid of all that Aristotelian nonsense in science. Why bring it back? But then, many of the central ideas of modern science were once considered preposterous. Consider, to name only three, heliocentrism, held by Aristarchus; atomism, held by Democritus; and evolution, held by Lamarck, and before him by even more disreputable characters. Of course, the modern versions of those ideas are quite different. But the fact of the matter is that thinkers like Copernicus,

Dalton, and Darwin found promise in those discredited ideas and took the trouble to develop them. To those thinkers we owe in large part the glory of modern science. Here we find in action both the principles of proliferation and of tenacity.

The operation of these last two principles makes science appear far more "sloppy" and "irrational" than its methodological image. But as we have seen, the attempt to make science conform to that methodological image, the attempt, that is, "to make science more 'rational' and more precise is bound to wipe it out."[24] For, as Feyerabend argues, "what appears as 'sloppiness', 'chaos' or opportunism when compared with such [image] has a most important function in the development of those very theories which we today regard as essential parts of our knowledge of nature."[25] But then from the methodologist's point of view nothing can be ruled out, *anything goes*. And if methodology is equated with reason, science is and must be an irrational enterprise.

As Einstein once put it: "The external conditions which are set for [the scientist] by the facts of experience do not permit him to let himself be too much restricted, in the construction of his conceptual world, by the adherence to an epistemological system. He therefore must appear to the systematic epistemologist as a type of unscrupulous opportunist. . . ."[26]

The situation is then as follows. According to the rationalist, alias methodologist, alias systematic epistemologist, certain events in the history of science constitute progress. But for those events to come about some scientists have to be opportunistic enough to adopt "whatever procedure seems to fit the occasion." This means that even the best of methodological rules *must* be violated from time to time. But this inherent limitation of all rules implies that nothing can be ruled out once and for all. To a methodologist this amounts to an admission that *anything goes*. Therefore, *from the methodologist's point of view*, anarchy will be occasionally essential to science.

This principle of anarchy is the only principle that does not inhibit progress, according to Feyerabend. But can it really be the case that anything goes in science? Although Feyerabend sometimes points out how ideas that today are considered preposterous have much to offer, he does this partly for rhetorical purposes and partly to annoy his educated opponents. For he does not argue that all ideas and procedures fit all circumstances equally well. On the contrary, in the case of Galileo and others he illustrates how some specific ideas and procedures were particularly helpful. *Anything goes* not from his point of view, but from the point of view of one who thinks that only certain ideas or procedures are admissible. Anarchy is thus in the eye of the rationalist beholder, a point that Feyerabend expresses with emphasis: "...'*anything goes*' *does not express any conviction of mine, it is a jocular summary of the predicament of the rationalist.*"

If anarchy means ignoring the rationalist's rules from time to time, then science requires anarchy. This is not to say that the rules never apply. If anything goes, reason sometimes goes too. Nor is this simply a bromide to the effect that since science is a human activity it *cannot* be perfectly rational. The point is rather that science *must not* be perfectly rational (in the rationalist sense that equates rationality with adherence to rules), if it is to achieve progress.

Nor is Feyerabend putting forth a new methodology (e.g., a counter-inductive methodology such as "advance theories inconsistent with the facts"). What we cannot do is precisely to specify in advance whether the inductive or the counter-inductive rules will apply. This would not change even if somehow we could foresee the context that the scientist will face. For the different choices that he may make when faced with a new context may themselves change that context in many different ways. When Dalton imported into chemistry his notions of simple ratios, he changed as a consequence the concepts of mixture and compound, as well as the standards of chemical explanation.

It is nevertheless uncanny that Feyerabend seems to be blessed with such special knowledge about what is good and what is bad in science. He certainly seems to have no qualms about using words like "progress," "advance," "improvement," and so on. Is there perhaps at least a rational way of appraising episodes in the history of science? But Feyerabend does not have to figure out whether some episode or other *really* constituted progress. His concern is of an altogether different sort. As he says

> *Everyone can read the terms in his own way* and in accordance with the tradition to which he belongs. Thus for an empiricist, 'progress' will mean transition to a theory that provides direct empirical tests for most of its basic assumptions. Some people believe the quantum theory to be a theory of this kind. For others, 'progress' may mean unification and harmony, perhaps even at the expense of empirical adequacy. This is how Einstein viewed the general theory of relativity. *And my thesis is that anarchism helps to achieve progress in any one of the senses one cares to choose.* Even a law-and-order science will succeed only if anarchistic moves are occasionally allowed to take place.[27]

Has Feyerebend, then, done away with rationality in science? My own inclination, which I have followed elsewhere,[28] is to take to heart Kuhn's advice:

> ... If history or any other empirical discipline leads us to believe that the development of science depends essentially on behavior that we have thought to be irrational, then we should conclude not that science is irrational, but that our notion of rationality needs adjustment here and there.[29]

Those adjustments, given Feyerabend's results, are bound to be quite dramatic. It is difficult to see how philosophy of science could return to the days before *Against Method*.

3. Incommensurability

Feyerabend and Kuhn's claim that in times of radical scientific change experience may change, as we saw in the case of Galileo, had many other profound consequences in the philosophy of science. If this claim is correct, the growth of science is not cumulative, for there is a sense in which science may begin anew in times of revolution. This result was met with dismay by many philosophers of science. According to these philosophers, rationality in science requires that changes of points of view may be justified. And this justification must be made by showing that the winner is superior to the loser. The only "scientific" way to show this superiority was by comparing the views in question with experience: The "facts" were the common measuring standards by which the worth of theories was to be determined. But if we may have different sets of facts before and after the change, we no longer can rely on common standards, and thus we cannot show that the new view is superior to the old. Therefore if Kuhn and Feyerabend are correct it cannot be shown that science is a rational enterprise. This is the problem created by the thesis of *incommensurability*, which strictly speaking means "lack of a common measure."

It is easy to see why this problem comes very straightforwardly out of Feyerabend's analysis of the history of science, and we can also see why it should become a headache for an empiricist epistemology. Analytic philosophers, however, tended to see this problem in linguistic terms. Such a linguistic turn was not unexpected. Given that it was common to think of theories as sets of (theoretical) sentences, or theoretical "languages" of some sort, science was expected to grow by linguistic accumulation. If Newton's theory was replaced by Einstein's, then it must be subsumed and explained under Einstein's. And to say that Einstein's theory explains Newton's is to say that Newton's must be derivable from Einstein's. Explanation was logical reduction, and of course logical reduction requires logical derivation. But unfortunately logical derivation fails if the common terms change meaning from the premises (the explanans) to the conclusion (the explanandum), which was exactly what Feyerabend also claimed.[30]

Analytic philosophers also feared that the failure of logical derivation would amount to a failure of communication of the following sort: According to Feyerabend (and Kuhn as well), the meaning of scientific terms is holistic, that is, it depends on the entire theory in which it plays a role. When the theory changes, the meaning of the terms employed changes also. Across theory change, then, scientists are speaking different languages. From this it supposedly follows that scientists who hold competing points of view (old vs. new Kuhnian paradigms, for example) will fail to agree because they are bound to misunderstand each other.[31]

To be fair, Kuhn did speak of how practitioners of the old paradigm might not be able to see the "evidence" presented in support of an alternative as evidence at all,[32] and Feyerabend did speak of holistic meaning. Also, both philosophers explain how communication was possible across theoretical boundaries (you could learn both "languages," for example).

Nonetheless, my purpose in this section is to argue that the problem of incommensurability exists quite apart from any one particular theory of meaning. Feyerabend's holistic theory, the target of so many lectures and publications, is simply irrelevant, as he himself pointed out. In fact, he belittled the entire linguistic discussion:

> ... Putnam creates the impression that I am mainly interested in meanings and that I am eager to find change where others see stability. This is not so. As far as I am concerned, even the most detailed conversation about meanings belongs in the gossip columns and have no place in the theory of knowledge. This is true even in those cases where meanings are invoked to force a decision about some different matter. For even here their only function is to conceal some dogmatic statement which would not be accepted, if presented by itself, and without the chatter of semantic discussion.[33]

I will begin with a discussion of whether the transition from Newton to Einstein was cumulative, continue with an account of how the question of meaning crept in, and finish with an argument for the irrelevance of (non-trivial) semantic theories.

On the grounds of economy, let me deal with the first point by reminding the reader of Kuhn's remarks concerning the transition Newton-Einstein. According to Kuhn, "Einstein's theory can be accepted only with the recognition that Newton's was wrong."[34] But Newton's physics is widely thought to have a limited validity and is in fact still much in use by engineers and even many scientists. What Einstein did, according to Kuhn's critics, was to draw the proper boundaries for classical mechanics. Within those boundaries Newton's theory is essentially correct. Furthermore, taken within those boundaries (e.g., low velocities) it can be shown to be derived from relativistic mechanics. Therefore Kuhn is wrong: this century's revolution in physics did not replace what preceded it but rather added to it. Knowledge in physics has grown by accumulation.

This objection seems odd in light of the many baffling aspects of Einstein's special theory of relativity (to say nothing of his general theory). It is true, of course, that the two theories yield similar numerical values (although not the same) for mass, length, etc., at low velocities.

Nevertheless, when the critics say, as they are obliged to, that Newtonians were unscientific if they thought that Newtonian theory applied at velocities close to that of light, those critics forget that "the price of significant scientific advance is a commitment that runs the risk of being wrong."[35] Such a restriction, Kuhn points out, "forbids the scientist to rely upon a theory in his own research whenever that research enters an area or seeks a degree of precision for which past practice with the theory offers no precedent."[36] The result of accepting it "would be the end of the research through which science may develop further."[37] As Kuhn argues, to save Newtonian physics in this manner (from being overthrown) would permit us to save practically any theory that has been held by serious scientists, including the notorious phlogiston theory, for within the limits of its validity it is still valid (of course).

Moreover, the alleged derivation is spurious. A necessary condition for the validity of any derivation is that the terms preserve their meanings throughout (otherwise we commit the logical fallacy of equivocation). Both classical and relativistic mechanics employ concepts of space, time, mass, etc. But according to Kuhn, "The physical referents of [the] Einsteinian are by no means identical with those of the Newtonian concepts that bear the same name."[38] For example, Newtonian mass is conserved, Einsteinian is convertible with energy.

In a similar fashion Feyerabend also questions the attempt to identify classical mass with relative *rest* mass. And relativistic length, Feyerabend adds, "involves an element that is absent from the classical concept and is in principle excluded from it. It involves the *relative velocity* of the object concerned in some reference system." He concludes that "different magnitudes based on different concepts may give identical values on their respective scales without ceasing to be different magnitudes." Where the terms involved have different meanings, then, no derivation is possible. Since this situation will occur in all those cases where a new comprehensive view of the world is born out of conflict, scientific revolutions are marked by the incommensurability of the theories (or paradigms) involved.

Some critics have suggested that it is possible to provide a "common dictionary," so it is possible to compare, say, the predictive successes of two competing views of the world. And there is a clear sense in which that is indeed possible, but this acknowledgment does not affect Feyerabend's view. In fact, the present example provides an illustration. The laws that give such good agreement in numbers with Einstein's at low velocities are not really Newton's laws, as he would have understood them, but rather a relativistic version of them. They are Newton's laws transposed into Einstein's special theory of relativity with limits and parameters on them that would have been inconceivable to Newton himself. If we were to take laws, and Einstein's as well, as mere instruments for the management of data, the meanings of the crucial terms might remain the same. But those terms acquire meaning for the scientist only as his training makes it second nature for him to attach them to the world in ways guided by the view of the world accepted by his discipline. As long as the way a term is used has nothing to do with its meaning, then we do not have this semantic problem of incommensurability. On the other hand, if we do think that the use of the term has something to do with its meaning, then we must consider how our view of the world influences that use.

Even this much commitment to "holism" can be eschewed by Feyerabend. There were at the time two recurrent themes in analytic philosophy of science with respect to the meaning of scientific terms. At one extreme was Bridgman's operationism, according to which the meanings of scientific terms should be determined by operations (e.g., length is defined by specified ways of measuring: by the use of rods, by triangulations, or by bouncing radar signals off an object).[39] At the other extreme was the view, favored by Hempel, that scientific terms have what he called "theoretic import," which means roughly that the meaning of a term depends in part on the role it plays in a theory — meaning is therefore affected by its relations to other terms also used in the theory.[40] Whatever extreme we tend to favor, or a suitable combination of both, we cannot avoid incommensurability. When we change paradigms (to stick to Kuhn's terminology) we obviously may change the relevant "operations," for the new paradigm may offer new and different experimental and instrumental commitments. And if we are inclined towards Hempel's view, it seems that drastic theoretical change may surely lead to change in meaning as well.

There was also a fixation in some circles that the meaning of "theoretical terms" had to be determined by prior "observational terms" (this is inductivism imported into semantics, i.e., the equivalent of "theories have to be derived from the facts"). If this view were correct, the meaning of observational terms would be prior and thus could not be altered by changes in the meaning of theoretical terms. It seems, however, that among the many assumptions made in such an argument, most of them very questionable, special prominence must be given to the distinction between theory and observation. But if this view of meaning needs such a distinction to even get off the ground, to advance it against those who question the distinction is to beg the question. Furthermore, we have seen how Feyerabend's analysis of the history of science demolishes such a distinction.

It is very important that we understand how far the analytic philosophers' accounts are from Feyerabend's notion of incommensurability. "Mere difference of meanings," he says, "does not yet lead to incommensurability in my sense."[41] Feyerabend's point was that Einstein's *use* of the terms mass and length precludes Newton's. According to Feyerabend, to say that Einstein's and Newton's theories (or paradigms) are incommensurable is to say that the principles for constructing concepts in one *suspend* the principles of the other. That is, the relativistic ideas used to understand mass, length, and time do not permit the use of the alternative Newtonian ideas. And surely, since concepts like "mass," "length" and "time" are used to determine what counts as facts in physics, we may conclude that presenting Einstein's facts means suspending principles assumed in the construction of Newtonian facts.[42]

This last way of putting the matter can bring us out of an issue over which philosophers have burned up much slobber for little profit. Notice how it returns us to the starting point of our discussion of this issue: Incommensurability simply means that there are no common standards of measurement, i.e., that there may be no common sets of facts to judge one theory or paradigm superior to another. If universal principles in a theory suspend the principles (of concept formation) in a second, we realize that the empirical basis of the first is different from that of the second. But when we put the matter this way, we realize that we are just talking about the possibility of overthrowing the empirical basis. And this we have already observed in Section 1 without semantic mirrors.

Although critics often concentrate their attacks on what they call Kuhn and Feyerabend's "holistic theory of meaning," neither Kuhn nor Feyerabend need commit themselves to any particular theory of meaning beyond the trivial point that the meaning of scientific terms is *somehow* connected with how scientists use them. The problem of incommensurability is an epistemological consequence of their analysis of the history of science, not of semantical idiosyncrasies on their part. Of course, the meaning of scientific terms does change from time to time, but that is no longer of great philosophical importance. It was so once, when analytic philosophy of science favored the view that logical derivation was an essential component of scientific explanation and that logical reduction was the right model for the growth of scientific knowledge. Feyerabend put the matter in perspective thus:

> I should add that incommensurability is a difficulty for philosophers, not for scientists. Philosophers insist on stability of meaning throughout an argument while scientists, being aware that 'speaking a language or explaining a situation means both following rules and changing them' . . . are experts in the art of arguing across lines which philosophers regard as insuperable boundaries of discourse.[43]

4. Relativism

I do not mean to deny that an air of paradox surrounds the issue of incommensurability and that a rather distinct odor of skepticism clings to it. If our paradigm or basic theory determines, among other things, the constituents of the world, one may say that "the world changes when basic theory changes." Or as Kuhn puts it: "after a revolution scientists are responding to a different world." [44] "It is rather," he adds, "as if the professional community had been suddenly transported to another planet where familiar objects are seen in a different light and are joined by unfamiliar ones as well."[45]

There are some who find this an exciting discovery about the nature of science. But others feel that we are confronted with a new skeptical problem: If the furniture of the world is relative to our world view, then a revolution in science may well bring about a change of furniture. This is a prospect that most philosophers would find repugnant. Many of them did argue that even if scientific terms change meanings the important thing is that they still refer to the same objects. But on this point Feyerabend can be most irksome:

> we certainly cannot assume that two incommensurable theories deal with one and the same objective state of affairs (to make the assumption we would have to assume that both at least refer to the same objective situation. But how can we assert that 'they both' refer to the same situation when 'they both' never make sense together? Besides, statements about what does and what does not refer can be checked only if the things referred to are described properly, but then our problem arises again with renewed force.) Hence, unless we want to assume that they deal with nothing at all we must admit that they deal with different worlds and that the change (from one world to another) has been brought about by a switch from one theory to another.[46]

For example, it is difficult to deny that, as Kuhn so thoroughly illustrates, the terms "element" and "mixture" have referred to different objects at different times in the history of chemistry. As for the furniture of the world, Feyerabend points out, our epistemic activities have made gods disappear and replaced them with heaps of atoms in empty space. Such talk has riled the many who know in their bones that the world has not changed even though they cannot prove it. For them this relativism is the most obnoxious consequence of the problem of incommensurability.

Some years ago I would have discussed the arguments of Putnam, Scheffler, Shapere, and others who insisted in distinguishing the meaning of a term from its reference as a way to solve this problem of incommensurability.[47] I would have pointed out, as Feyerabend did, that since Bohr's analysis of EPR we know that "there are changes which are not the results of a causal interaction between object and observer but a change of the very conditions that permit us to speak of objects, situations, events."[48] I would have also pointed out, as I did elsewhere, that this "ontological" problem of incommensurability disappears within a sophisticated relativism:

> The issue of whether the world changes when we change basic theory does not arise within the relativity of science. To conceive of the world is to conceive of it within a frame of reference. To ask whether the world really changes when we change frames of reference is comparable to asking (in the Special Theory of Relativity) whether the mass really changes when we change frames of reference. It is just not a sensible question.[49]

The pursuit of these issues, however, does not fit the purposes of this paper. As it turns out, Feyerabend himself became a critic of relativism in later years. This drastic change in his position, as we will see shortly, simply introduces into his treatment of the issue of reality one of the main themes of his philosophy of science:

> A theory (model, sketch) of a historical process says both too little and too much. It says too little because it starts from a fragment of what it wants to represent. But it also says too much because the fragment is not just described, it is subdivided into essence and accident, the essence is generalised and used to judge the rest."[50]

Relativism proves unsatisfactory, then, because it is a *theory* of a historical process, knowledge. Relativists, Feyerabend believes, accept the assumption that "the theories, facts, and procedures that constitute the (scientific) knowledge of a particular time are the results of specific and highly idiosyncratic historical developments."[51] At the same time they reject the assumption that "what has been found in [an] idiosyncratic and culture-dependant way... *exists* independently of the circumstances of its discovery."[52] They hold, for instance, that "atoms exist *given* the conceptual framework that projects them."[53] The trouble with relativism, Feyerabend argues, is that

> ...traditions not only have no well-defined boundaries, but contain ambiguities and methods of change which enable their members to think and act as if no boundaries existed: potentially every tradition is all traditions. Relativizing existence to a single "conceptual system" that is then closed off from the rest and presented in unambiguous detail mutilate real traditions and creates a chimera.[54]

This approach echoes Feyerabend's approach to theories of scientific rationality: every one of them, too, circumscribes the practice of science to a single system (of methodological norms or standards) which is then closed off from the rest (in that it is the only acceptable one). But standards,

Feyerabend says, "are intellectual measuring intruments; they give us readings not of temperature, or of weight, but of the properties of complex sections of the historical process."[55] To presume that we can apply such standards universally to the practice of science is similar to supposing that we can satisfactorily answer "the question what measuring intruments will help us to explore an as yet unspecified region of the universe. We don't know the region, we cannot say what will work in it. . . (as an example consider the question how to measure the temperature in the center of the sun, put at about 1820)."[56] Likewise, relativizing existence to a single conceptual system means relativizing it to a single intellectual tradition, and this requires that we ignore how traditions may change, borrow from others, adapt to new circumstances; in short, it means trying to force an abstract "time-slice" on an open-ended historical process.

Realism, at least in its most common varieties, would fare no better in this regard:

> Both objectivism (and the associated idea of truth) and relativism assume limits that are not found in practice and postulate nonsense wherever people are engaged in interesting though occasionally difficult forms of collaboration. Objectivism and relativism are chimeras.[57]

Thus Feyerabend came to believe in the need for a new approach to the problem. He pointed out with the relativists that in auspicious circumstances the entities we project "do indeed 'appear' in a clear and decisive way." [58] But then he reminded us, with Bohr, that such appearances may be regarded as phenomena "that transcend the dichotomoy subjective/objective,"[59] which underlies the clash between realism and relativism. "They are 'subjective,'" he says, "for they could not exist wihtout the idiosyncratic conceptual and perceptual guidance of some point of view. . . But they are also 'objective': not all ways of thinking have results and not all perceptions are trustworthy."[60] This is so, because the material humans face "offers resistance; some constructions . . . find no point of attack in it and simply collapse."[61]

In several articles that Feyerabend wrote towards the end of his life, he elaborated the same theme. Quantum theory, one of the best confirmed theories we have, he reminded us, "implies, in a widely accepted interpretation, that properties once regarded as objective depend on the way in which the world is being approached."[62] Nevertheless, he warned, "at this point it is important not to fall into the trap of relativism,"[63] for as he had just pointed out, *"nature as described by our scientists is indeed an artifact,"* but an artifact *"built in collaboration with a Being sufficiently complex to mock and, perhaps, punish materialists by responding to them in a crudely materialistic way."* [64]

Nature as described by scientists is not "Nature In And For Herself, it is the result of an interaction between two rather unequal partners, tiny men and women on the one side and Majestic Being on the other."[65] This emphasis on such elusive "Being," I believe, marks the return of Feyerabend to some sort of realism, although clearly not to any traditional form or realism, and not quite to an internal realism *a la* Putnam either. It comes closer, perhaps, to C.A. Hooker's evolutionary realism, in that it uses evolutionary motifs and insists on pluralism (a Feyerabend influence on Hooker). Not all interactions, says Feyerabend, produce beneficial results:

> Like unfit mutations, the actors of some exchanges (the members of some cultures) linger for a while and then disappear (different cultures have much in common with different mutations living in different ecological niches). The point is that there is no only one successful culture, there are many and that their success is a matter of empirical record, not of philosophical definitions...[66]

Feyerabend is not limited to pointing out that "the world is much more slippery than is assumed by our rationalists," for "there is also a positive result, namely, an insight into the abundance that surrounds us and that is often concealed by the imposition of simpleminded ideologies."[67] It is true, he says,

> ...that allowing abundance to take over would be the end of life and existence as we know it — abundance and chaos are different aspects of one and the same world. We need simplifications (e.g., we need bodies with restricted motions and brains with restricted modes of perception). But there are many such simplifications, not just one and they can be changed to remove the elitism which so far has dominated Western Civilization.[68]

I have argued elsewhere for an evolutionary relativism that offers the flexibility that Feyerabend demands and that fits in well with the open-ended historicity of knowledge that he so ably defends.[69] Thus I do not believe that in order to transcend the common dichotomy realism/relativism it is necessary to make references to an ineffable Being. Others may disagree with different aspects of Feyerabend's complex view on the issue of relativism (which changed considerably after his sophisticated defense of a practical relativism in *Farewell to Reason*).[70] But we should all agree that his latest view enlightens us about some important hurdles that await any theory of knowledge. Far from holding some whimsical and (easily?) refutable doctrine, Feyerabend has pushed time and again the frontiers of the philosophy of science.

5. Conclusion

I hope I have been able to convey to the reader some of the complexity and the richness of Paul Feyerabend's epistemology of science, some of the depth of his analysis of the practice of science, and some of the principled ways in which the problems he uncovered, and with which he struggled throughout his career, are consequences of that analysis. Of course, his philosophy went well beyond the issues to which I have alluded in this paper, but I must content myself with these approximations and simplifications of a body of thought that in itself exemplified the open-ended character of knowledge.

Notes

1. Paul K. Feyerabend, "Philosophy of Science Versus Scientific Practice: Observations on Mach, his Followers and his opponents." *Philosophical Papers. Vol. 2: Problems of Empiricism*. Cambridge University Press. 1981. p. 80.
2. This analysis reached its fullest expression in Feyerabend's *Against Method*. NLB. 1975.
3. This particular point is developed in detail in *Against Method, op. cit.*, Ch. 6, pp. 69-80. The account of the Tower Argument appears in Chapters 6-8, and the account of the telescope follows it in Chapters 9-11 (9-10 in the third edition).
4. See, for example, R. Rorty, *Philosophy and the Mirror of Nature*, Princeton University Press, 1979, p. 246.
5. *Ibid.*, p. 101. The original quotation and the ones that follow below are from Galileo Galilei, *Dialogue Concerning the Two Chief World Systems*, University of California Press, 1953 (appropriate references in *Against Method*).
6. *Ibid.*
7. *Ibid.*, p. 103.
8. *Ibid.*, p. 124.
9. In Galileo's time the concern would have been about the transition from the superlunary to the sublunary region.
10. It is wise to take such stories with a grain of salt. Nevertheless, Einstein did write to Born, concerning the latter's remark that Freundlich's analysis of the bending of light near the sun and of the redshift showed that Einstein's formula was not quite right, "Freundlich. . . does not move me in the slightest. Even if the deflection of light, the perihelial movement or line shift were unknown, the gravitation equations would still be convincing because they avoid the inertial system (the phantom which affects everything but is itself not affected). *It is really strange that human beings are normally deaf to the strongest arguments while they are always inclined to overestimate meausuring accuracies.*" Quoted in *AM*, p 57n (Feyerabend's emphasis).
11. *Ibid.*, pp. 32-33.
12. *Science in a Free Society,* NLB, 1978, pp. 142-143.
13. *Ibid.*, p. 143.
14. *Ibid.*

15. T.S. Kuhn, *The Structure of Scientific Revolutions* (2nd. edition), The University of Chicago Press, p. 78.
16. *Ibid.*
17. *Ibid.*
18. *AM*, p. 18.
19. *Ibid.*, p. 17.
20. *Ibid.*
21. T.S. Kuhn, *op. cit.*, pp. 23-24 and 157-158.
22. This account is based principally on Feyerabend's famous article "Consolations for the Specialist," which first appeared in I. Lakatos and A. Musgrave (eds.), *Criticism and the Growth of Knowledge*, Cambridge University Press, 1970.
23. *AM*, p. 32.
24. *Ibid.*, p. 179.
25. *Ibid.*
26. *Albert Einstein: Philosopher-Scientist*, P.A. Schilpp (ed.), Open Court (third edition), 1982, p. 684.
27. *AM*, p. 27.
28. See for example *Radical Knowledge*, Hackett Publishing Co., 1981, and "Evolution and Justification," *The Monist*, Vol. 71, No. 3, July 1988, pp. 339-357. In these and other essays I argue for a social conception of scientific rationality.
29. T.S. Kuhn, "Notes on Lakatos," in R.S. Cohen and R.C. Buck (eds.), *Boston Studies in the Philosophy of Science*, Vol. VIII, 1971, p. 144.
30. See Feyerabend's landmark article "Explanation, Reduction and Empiricism," *Minnesota Studies in the Philosophy of Science*, Vol. 3, 1962, now reprinted in his *Philosophical Papers*, op. cit., pp. 44-96.
31. Often a similar point was made on the basis of the impossibility of translation that incommesurability was supposed to imply. See, for example, H. Putnam, *Reason, Truth, and History*, Cambridge University Press, 1981, p. 114.
32. T.S. Kuhn, *op. cit.*, p. 94.
33. "Comments on Smart, Sellars and Putnam," *Philosophical Papers*, Vol. 1, Cambridge University Press, 1981, p. 113.
34. *Ibid.*, p. 98.
35. *Ibid.*, p. 101.
36. *Ibid.*, p. 100.
37. *Ibid.*
38. *Ibid.*, p. 102.
39. P. Bridgman, *The Logic of Modern Physics*, The Macmillan Co., 1927.
40. C. Hempel, *Philosophy of Natural Science*, Prentice Hall, 1966.
41. "Putnam on Incommensurability," *Farewell to Knowledge*, Verso, 1987, p. 272.
42. *AM*, pp. 267-285.
43. *Ibid.*
44. T.S. Kuhn, *The Structure of Scientific Revolutions*, op. cit., p.111.
45. *Ibid.*
46. *Science in a Free Society*, NLB, 1978, p. 70.
47. H. Putnam, *Meaning and the Moral Sciences*, Routledge and Kegan Paul, 1978; I. Scheffler, *Science and Subjectivity*, Bobbs-Merrill Co., 1967; D. Shapere, "Meaning and Scientific Change," in R. Colodny (ed.), *Mind and Cosmos*, University of Pittsburgh Press, 1966. For an interesting discussion see A.N.

Perovich, "Incommensurability, Its Varieties and Its Ontological Consequences," in G. Munevar (ed.) *Beyond Reason: Essays on the Philosophy of Paul K. Feyerabend*, Vol. 132 of *Boston Studies in the Philosophy of Science*, Kluwer Academic Publishers, 1991, pp. 313-328.
48. *Ibid.*
49. *Radical Knowledge*, Hackett, 1981, pp. 56-57.
50. "Realism," in C.C. Gould and R.S. Cohen (eds.), *Artifacts, Representations and Social Practice*, Kluwer Academic Publishers, 1994, p. 207.
51. "Realism and the Historicity of Knowledge," *The Journal of Philosophy*, Vol. LXXXVI, No. 8, 1989, p. 393.
52. *Ibid.*, p. 394.
53. *Ibid.*, p. 403.
54. *Ibid.*, pp. 404-405.
55. *Science in a Free Society, op. cit.*, p. 37.
56. *Ibid.*
57. "Potentially Every Culture is All Cultures," *Common Knowledge*, 1993, p. 20.
58. *Ibid.*, p. 404.
59. *Ibid.*
60. *Ibid.*
61. *Ibid.*, p. 405.
62. "Art as a Product of Nature as a Work of Art," *World Futures*, Vol. 40, p. 98.
63. *Ibid.*
64. *Ibid.*
65. *Ibid.*
66. *Ibid.*
67. *Ibid.*, p. 99.
68. *Ibid.*
69. See especially my "Evolution and the Naked Truth," in M. Dascal (ed.), *Cultural Relativism and Philosophy*, E.J. Brill, 1991, pp. 177-194.
70. "Notes on Relativism," *Farewell to Reason*, Verso, 1987, pp. 19-89.

Index

Alexander, R., 134, 147n
Alley, C., 186, 188n
Amado, J., 140, 142
anarchist, 204, 225
anarchy, 89n, 105, 121, 227, 229, 231, 232
Aristotle, 46, 58, 88n, 146, 201, 202, 203, 206, 209, 210, 221, 224, 225

Beck, L., 31n
biology, x, 35, 36, 39, 42, 47, 50, 66, 84, 91, 98n, 110n, 117, 124, 127, 128, 131, 135, 136, 139, 142, 144, 145, 147n, 161, 189, 216n, 223, 224
Blackwell, B., 62n, 147n, 166n
Bohr, N., ix, 33, 34, 39, 40, 42, 43n, 176, 238, 240
Boyd, R., 33, 43n
Bradie, M., 110n, 131, 134, 146, 165n, 166n
Bridgman, P., 236, 243n
Bronowski, J., 73n, 101
Brush, S., 188n

Campbell, D., 100
Carnap, R., 76, 79
Cederblom, J., 32n
Churchland, P., xi, 146, 147n, 150, 152, 153, 154, 155, 156, 161, 162, 165n, 166n, 195n
Clark, W., 172, 179n
Clarke, A., 169, 179n
convergence, 29, 32n, 38, 48, 146
Copernicus, 30, 88n, 174, 183, 184, 189, 212, 221, 222, 223, 230
Crick, F., 150, 159, 164n, 212
curiosity, 55n, 69, 71, 72n, 84, 95, 97, 101, 102, 103, 108, 109, 117, 118, 119, 121, 172, 176, 178

Darwin, xi, 132, 133, 134, 147n, 161, 164n
Darwinian, ix, x, xi, 121, 131, 143, 131, 134
Davies, P., 195n
Dawkins, R., 133
determinism, 149, 151, 152, 153, 155, 156, 157, 158, 159, 160, 161, 165n
Dreyfus, R., 195n
Duhem, P., 58

Einstein, A., 35, 40, 43n, 77, 152, 158, 175, 176, 186, 187, 224, 231, 232, 233, 234, 235, 237, 242n, 243n

empiricism, 35, 76, 113, 114, 210, 220, 222, 230, 242n, 243n
empiricists, 75, 76, 78, 85, 88n, 115, 222, 224
  logical empiricists, 75, 78, 115
epiphenomenalism, 154
epistemology, ix, x, xi, 35, 36, 39, 40, 58, 59, 60, 65, 66, 67, 75, 80, 83, 84, 85, 86, 88n, 89n, 91, 92, 93, 95, 97, 98n, 99, 100, 105, 106, 108, 110n, 113, 115, 116, 121, 123, 127, 129, 130n, 215, 216n, 219, 224, 226, 227, 233, 242n
  evolutionary epistemology, x, 65, 80, 91, 92, 100, 108, 116
ethics, x, 68, 108, 113, 123, 124, 127, 128, 129, 130n, 131, 133, 134, 135, 136, 137, 138, 139, 144, 145, 146, 147n, 166n
Everitt, F., 186, 188n
evolution, x, 27, 31n, 35, 47, 48, 65, 66, 67, 68, 69, 70, 72n, 83, 84, 91, 93, 94, 98n, 100, 106, 108, 113, 116, 117, 143, 144, 146, 163, 164n, 166n, 183, 184, 185, 190, 230, 243n, 244n
evolutionary, ix, x, xi, 29, 30, 33, 35, 39, 40, 41, 42, 43n, 49, 50, 53, 55n, 61, 65, 66, 67, 69, 72n, 73n, 80, 84, 85, 86, 87, 91, 92, 94, 98n, 99, 100, 101, 104, 106, 108, 110n, 116, 117, 119, 120, 121, 122, 131, 132, 134, 135, 143, 144, 145, 146, 147n, 149, 161, 223, 224, 241
extraterrestrial, x, 23, 31n, 189, 195n

Feyerabend, P., ix, x, xi, 40, 43n, 57, 58, 60, 67, 68, 72n, 75, 77, 78, 79, 80, 81, 82, 83, 85, 88n, 89n, 90n, 99, 105, 110n, 114, 115, 121, 130n, 173, 174, 179n, 199, 200, 201, 202, 203, 204, 205, 206, 207, 208, 209, 210, 211, 212, 213, 214, 215, 216n, 217n, 219, 220, 221, 222, 224, 225, 226, 227, 228, 229, 230, 231, 232, 233, 234, 235, 236, 237, 238, 239, 240, 241, 242n, 243n, 244n
Folse, H., 43n
free will, xi, 146, 149, 150, 151, 152, 153, 154, 155, 156, 157, 158, 159, 161, 162, 164n, 165n, 166n
Frege, G., 79

Galileo, 77, 88n, 114, 174, 175, 183, 184, 188n, 205, 210, 212, 220, 221, 222, 223, 224, 225, 231, 233, 242n
general relativity, 152, 186, 187, 188n, 224
Goodman, N., 120
Gould, S., 32n, 244n
Gregory, R., 69
Greve, T., 169, 179n
Grush, R., 155, 165n

Hahlweg, K., 95, 98n
Hardin, G., 133
Heisenberg, W., 34, 152, 176
Hempel, C., 75, 76, 78, 79, 85, 88n, 90n, 113, 115, 130n, 236, 243n
Hobbes, 157, 159
Holman, M., 171, 179n
Hooker, C., 38, 43n, 59, 60, 62n, 95, 98n, 106, 110n, 241
Horwich, P., 165n
Hull, D., x, 91, 92, 93, 94, 96, 97, 98n, 106, 110n
Hume, D., xi, 124, 131, 137, 146, 149, 156, 157, 158, 159, 164n, 166n

incommensurability, 59, 206, 207, 219, 220, 233, 234, 235, 236, 237, 238, 243n, 244n
indeterminism, 149, 153, 154, 155, 156, 166n

James, W., 153, 157, 158, 165n

Kant, 31n, 95
Kline, M., 89n
Kuhn, T., x, 42, 57, 58, 60, 68, 70, 72n, 75, 77, 78, 79, 81, 83, 85, 89n, 99, 103, 110n, 114, 130n, 174, 179n, 203, 204, 215, 224, 226, 227, 228, 229, 230, 232, 233, 234, 235, 236, 237, 238, 243n

Lakatos, I., 68, 72n, 76, 77, 78, 79, 80, 81, 82, 83, 86, 88n, 89n, 90n, 93, 99, 107, 110n, 115, 121, 123, 130n, 174, 179n, 200, 203, 215, 243n
Lewis, C., 31n
Lied, F., 179n
Livingston, R., 72n
Lorenz, K., x, 65, 66, 71, 101, 110n
Luria, S., 164n

Mach, E., 65, 66, 67, 72n, 84, 219, 242n
Margolis, J., x, 57, 58, 59, 60, 62n
Margulis, L., 24, 31n
Matson, W., ix
Maxwell, C., 30, 173
Mill, J., 157, 159, 163, 164n, 166n, 214, 217n, 230
Mosterín, J., 45, 55n

Munévar, G., 43n, 55n, 62n, 94, 96, 97, 98n
Musgrave, A., 72n, 73n, 110n, 243n

Nansen, F., 169
NASA, xi, 187, 188n, 190, 195n
naturalism, x, 80, 81, 83, 85, 97, 113, 116, 128, 129, 149, 155, 160, 164n, 165n, 202
naturalist, 25, 83, 149, 150, 164n, 203
naturalistic, x, xi, 35, 47, 80, 91, 99, 106, 108, 113, 116, 122, 123, 124, 128, 130n, 136, 139, 143, 147n, 149, 150, 155, 156, 160, 161, 169
  naturalistic fallacy, x, 124, 136, 139, 143, 147n, 169
naturalized ethics, 145
Newton, 27, 28, 30, 184, 233, 234, 235, 237

Ortega y Gasset, 45

Penrose, R., 155, 165n
Piaget, J., 25, 68, 100, 110n, 117, 130n, 216n
Plato, 40, 45, 49, 116, 122, 133, 142
Poincare, H., 65, 66, 165n
Polanyi, M., 80, 81, 203
Popper, K., ix, 33, 40, 43n, 65, 67, 71, 83, 100, 108, 110n, 116, 130n, 174, 179n, 202
Prigogine, I., 155, 156, 161, 165n, 166n
principle of mediocrity, 189
Putnam, H., 59, 234, 238, 241, 243n

quantum physics, ix, 33, 34, 152, 153, 154, 155, 176
Quine, W., x, 58, 116, 130n

rationality, x, xi, 68, 75, 76, 77, 80, 81, 85, 86, 87, 88n, 89n, 91, 93, 94, 95, 96, 99, 103, 105, 107, 108, 109, 113, 114, 116, 119, 120, 121, 122, 123, 130n, 144, 187, 200, 201, 202, 204, 206, 209, 226, 232, 233, 239, 243n
Rawls, J., 128, 129, 130n, 138, 139, 144, 145, 147n
realism, ix, 33, 34, 35, 37, 40, 41, 43n, 47, 54, 57, 59, 60, 61, 62n, 65, 101, 108, 207, 240, 241, 244n
Reid, 156
relative truth, ix, 42, 47, 52, 53, 54, 55n, 145
relativism, ix, x, 33, 40, 41, 49, 55n, 45, 46, 47, 49, 50, 53, 57, 58, 59, 60, 87, 108, 131, 143, 144, 147n, 199, 201, 202, 203, 204, 206, 207, 215, 216n, 219, 238, 239, 240, 241, 244n
  evolutionary relativism, ix, xi, 40, 41, 42, 49, 55n, 61, 131, 145, 241
  social relativism, 3, 4, 19
Rescher, N., 31n

Index 247

rhetoric, 181, 187, 199
Richards, R., 134, 147n
Rorty, R., 75, 86, 87, 88n, 90n, 242n
Russell, B., 79

Sagan, C., 31n, 184, 195n
Schlick, M., 149, 158, 159, 162, 164n, 166n
scientific knowledge, ix, 65, 66, 68, 70, 72n, 76, 80, 84, 87, 96, 116, 173, 178, 237
Scriven, M., ix
self-determination, 156, 161
Sellars, W., 86, 87, 243n
serendipity, 172, 173, 176, 177, 178
SETI, 23, 29, 31n, 32n, 189
Singer, P., x, 110n, 124, 125, 127, 128, 129, 130n, 131, 132, 133, 134, 135, 136, 137, 138, 139, 141, 142, 143, 144, 145, 147n
Sluga, H., ix

Spencer, H., 65
Stent, G., ix

Tandberg, E., 179n
Tipler, F., 195n
Toulmin, S., 43n, 67, 80, 81, 88n, 89n, 100, 106, 110n, 116, 121, 130n

universalism, 45, 47, 53, 55n

Von Neumann, J., xi, 190, 192, 193, 195n

Watson, G., 156, 159, 160, 162, 166n, 212
Wilson, E., x, 124, 127, 129, 131, 135, 136, 137, 139, 140, 141, 142, 143, 144, 145, 147n, 151, 152, 153, 162, 165n, 172, 179n
Wittgenstein, 86, 88n, 130n, 152, 217n

BD 177 .M85 1998
Munevar, Gonzalo.
Evolution and the naked
 truth